CW01603461

This book is to be ret... ...before
the last date stamped below.

20. APR. 1993

WITHDRAWN

107402

Butterworths Technical and Scientific Checkbooks

Engineering Mathematics and Science 3 Checkbook

J O Bird
BSc, AFIMA, TEng(CEI), MITE

A J C May
BA, CEng, MIMechE, FITE, MBIM

D S Ayling
BSc(Eng), CEng, MIMechE, MRINA

Butterworths
London Boston Sydney Wellington Durban Toronto

All rights reserved. No part of this publication may be reproduced or transmitted in any form or by any means, including photocopying and recording without the written permission of the copyright holder, application for which should be addressed to the publishers. Such written permission must also be obtained before any part of this publication is stored in a retrieval system of any nature.

This book is sold subject to the Standard Conditions of Sale of Net Books and may not be resold in the UK below the net price given by the Publishers in their current price list.

First published 1982

© Butterworth & Co (Publishers) Ltd 1982

British Library Cataloguing in Publication Data

Bird, J.O.
Engineering mathematics & science 3 checkbook.
1. Shop mathematics
I. Title II. May, A.J.C. III. Ayling, D.S.

ISBN 0-408-00670-6
ISBN 0-408-00625-0 Pbk

Typeset by Scribe Design, Gillingham, Kent
Printed by Hartnoll Print Ltd, Bodmin, Cornwall

Contents

Note to Reader

As textbooks become more expensive, authors are often asked to reduce the number of worked and unworked problems, examples and case studies. This may reduce costs, but it can be at the expense of practical work which gives point to the theory.

Checkbooks if anything lean the other way. They let problem-solving establish and exemplify the theory contained in technician syllabuses. The Checkbook reader can gain *real* understanding through seeing problems solved and through solving problems himself.

Checkbooks do not supplant fuller textbooks, but rather supplement them with an alternative emphasis and an ample provision of worked and unworked problems. The brief outline of essential data—definitions, formulae, laws, regulations, codes of practice, standards, conventions, procedures, etc—will be a useful introduction to a course and a valuable aid to revision. Short-answer and multi-choice problems are a valuable feature of many Checkbooks, together with conventional problems and answers.

Checkbook authors are carefully selected. Most are experienced and successful technical writers; all are experts in their own subjects; but a more important qualification still is their ability to demonstrate and teach the solution of problems in their particular branch of technology, mathematics or science.

Authors, General Editors and Publishers are partners in this major low-priced series whose essence is captured by the Checkbook symbol of a question or problem 'checked' by a tick for correct solution.

Preface

This textbook of worked problems provides coverage of the Technician Education Council's level 3 unit in Engineering Mathematics and Science (syllabus U76/065). However it can be regarded as a basic textbook in Engineering Mathematics and Science for a much wider range of courses.

The aim of the book is to further develop the studies of Science and Mathematics for students engaged in 'practical approach' programme and which will complement other units in the programme for General Engineering. Each topic considered in the text is presented in a way that assumes in the reader only the knowledge attained at TEC level 2 in Engineering Science and Mathematics. This practical book contains nearly 100 illustrations, some 180 detailed worked problems, followed by over 400 further problems with answers.

The authors would like to express their appreciation for the friendly co-operation and helpful advice given to them by the publishers. Thanks are also due to Mrs Elaine Mayo for the excellent typing of the manuscript.

Finally, the authors would like to add a special word of thanks to their wives, Elizabeth, Juliet and Joan for their patience, help and encouragement during the preparation of this book.

JO Bird
AJC May
DS Ayling

Highbury College of Technology
Portsmouth

Butterworths Technical and Scientific Checkbooks

General Editors for Science, Engineering and Mathematics titles:
J.O. Bird and A.J.C. May, Highbury College of Technology, Portsmouth.

General Editor for Building, Civil Engineering, Surveying and Architectural titles:
Colin R. Bassett, lately of Guildford County College of Technology.

A comprehensive range of Checkbooks will be available to cover the major syllabus areas of the TEC, SCOTEC and similar examining authorities. A comprehensive list is given below and classified according to levels.

Level 1 (Red covers)
Mathematics
Physical Science
Physics
Construction Drawing
Construction Technology
Microelectronic Systems
Engineering Drawing
Workshop Processes & Materials

Level 2 (Blue covers)
Mathematics
Chemistry
Physics
Building Science and Materials
Construction Technology
Electrical & Electronic Applications
Electrical & Electronic Principles
Electronics
Microelectronic Systems
Engineering Drawing
Engineering Science
Manufacturing Technology
Digital Techniques
Motor Vehicle Science

Level 3 (Yellow covers)
Mathematics
Chemistry
Building Measurement
Construction Technology
Environmental Science
Electrical Principles
Electronics
Microelectronic Systems
Electrical Science
Mechanical Science
Engineering Mathematics & Science
Engineering Science
Engineering Design
Manufacturing Technology
Motor Vehicle Science
Light Current Applications

Level 4 (Green covers)
Mathematics
Building Law
Building Services & Equipment
Construction Technology
Construction Site Studies
Concrete Technology
Economics for the Construction Industry
Geotechnics
Engineering Instrumentation & Control

Level 5
Building Services & Equipment
Construction Technology
Manufacturing Technology

1 Materials

A. MAIN POINTS CONCERNING TEMPERATURE STRESS, HARDNESS TESTS AND IMPACT TESTS

1 **Temperature stress**

(i) Unrestrained expansion or contraction due to change of temperature is given by $l \alpha t$,
where l = length;
α = coefficient of linear expansion;
t = change of temperature.
The coefficient of linear expansion, a constant for a given material, is a very small fraction per degree Celsius (°C); if t is in °C then the change of length is in the same units as the original length. (See *Problem 1*)

(ii) If expansion or contraction due to change of temperature is prevented or restricted, stress will be set up. If the bar in *Fig 1* for example is heated while it is held between rigid stops, forces F will be set up, producing stress, and therefore strain.

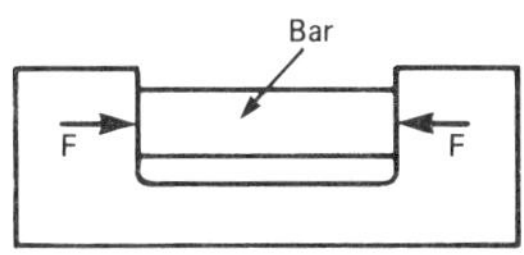

Fig 1

(iii) $$\text{Stress } (\sigma) = \frac{\text{force}}{\text{area}};$$

$$\text{Strain } (\epsilon) = \frac{\text{change of length}}{\text{original length}}$$

$$\frac{\text{Stress}}{\text{Strain}} = \text{Modulus of elasticity or Young's modulus (E)}$$

Strain is the ratio of two lengths and therefore has no units. The basic units of stress and modulus of elasticity are therefore the same, the pascal (Pa) equivalent to 1 N/m^2. This is a very small unit for this work and more convenient units are:

for stress, the megapascal (MPa) where 1 MPa = 1 $N/m^2 \times 10^6$
for modulus, the gigapascal (GPa) where 1 GPa = 1 $N/m^2 \times 10^9$
(Note. 1 MPa × 1 mm^2 = 1 N).

(iv) If expansion is completely prevented, as in *Fig 1*, the changes in length due to stress and temperature change balance:

$$l \times \text{strain} = l \alpha t$$

$$\text{i.e. } l \times \frac{\text{stress}}{E} = l \alpha t$$

The stress will be compressive stress. Tensile stress is set up if contraction due to fall in temperature is prevented.

(v) If expansion or contraction is restricted rather than prevented then $l\,\alpha\,t - \dfrac{l\,\sigma}{E}$ will equal the expansion or contraction permitted. Again tensile stress will be associated with fall in temperature and compressive stress with temperature rise. (See *Problems 2 to 4*)

2 Hardness and other non-destructive tests

(a) HARDNESS of a material can be defined in several ways, none of which is entirely satisfactory:

(i) the ability to scratch other materials;
(ii) the ability to resist scratching;
(iii) the ability to resist plastic spreading under indentation;
(iv) the ability to resist elastic deformation under indentation;
(v) the ability to resist deformation by rolling.

(b) THE BRINELL HARDNESS TEST

This test is performed by pressing a hardened steel ball into the given material under a specified load and then measuring the diameter of the indentation produced. The Brinell hardness number (H_B value) is obtained by dividing the applied load in kg by the spherical area of the indentation in mm^2.

$$\text{Brinell hardness number} = \frac{F}{\dfrac{\pi D}{2}\left[D - \sqrt{(D^2 - d^2)}\right]},$$

where

F = applied load (kg),
D = diameter of ball (mm)
d = diameter of indentation (mm)
(See *Fig 2*)

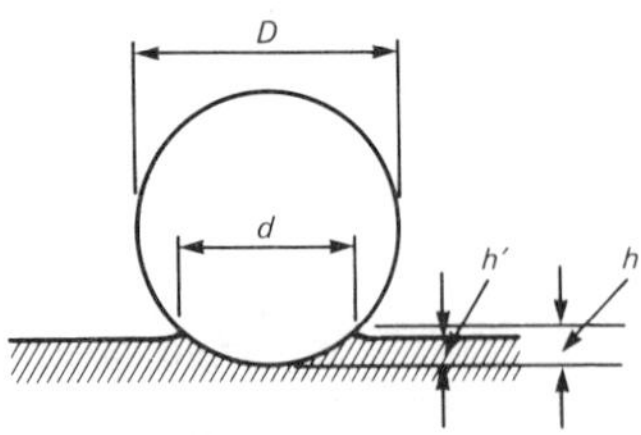

Fig 2

Although the spherical area is readily obtained from the depth of the impression, h in *Fig 2*, this is much more difficult to measure than h'. Also, strictly speaking, the area used should be that of the indentation under load. When the load is removed there is some elastic recovery, different for different materials, and this affects the depth of impression more significantly than the diameter of the impression.

For consistent and comparable results, the Brinell hardness test must be arranged to give a value of d/D in the range 0.25 to 0.5, and ideally 0.375. To achieve this the original arrangement of a 3000 kg load and a 10 mm ball has been extended to other loads and other sizes of ball. The original arrangement could only be used on large test pieces.

An acceptable value for d/D is ensured by using F/D^2 values specified for different materials by the British Standards Institution which has also standardised loads and ball sizes. Some specified F/D^2 values are:

Steels, cast iron	30
Copper and aluminium alloys	10
Copper, aluminium	5
Lead, tin	1

If an unidentified material is to be tested, values of F/D^2 must be experimented with until an acceptable value of d/D is obtained. The usual ball sizes available are 1 mm, 2 mm, 5 mm and 10 mm. Standard loads are as shown in *Table 1.*

TABLE 1

Ball diameter (mm)	*Load* (kg)			
	$F/D^2 = 1$	$F/D^2 = 5$	$F/D^2 = 10$	$F/D^2 = 30$
1	1	5	10	30
2	4	20	40	120
5	25	125	250	750
10	100	500	1000	3000

The Brinell test is unsuitable for very hard materials because of the deformation of the ball in achieving an acceptable d/D value. The ball flattens and gives a misleadingly high value of d, leading to an H_B value below the true one.

Hardness tests can be used to calculate approximate values of ultimate tensile strengths. The factor by which the Brinell hardness number should be multiplied to give, for steels for example, the u.t.s. in MPa, usually lies in the range 3 to 3.5. The figures do not apply to some special steels. The following precautions are required when carrying out a Brinell test.

(i) The material must be sufficiently wide and thick. The impression must have its centre not less than two and a half times its diameter from any edge. The thickness must be at least ten times the depth of the impression, this depth being given by $F/(\pi DH)$ where H is the estimated Brinell hardness number.

(ii) The surface of the material should, if possible, be ground flat and polished:

(iii) The load should be held for 15 s;

(iv) Two diameters of the impression at right angles should be read and their mean used in the calculation;

(v) When stating the result of the test the ball number and load should be stated, e.g. H 10/3000 = 410.

Special machines are made but a Brinell test can be carried out on most universal testing machines.

(c) THE VICKERS DIAMOND PYRAMID HARDNESS TEST

This test uses a diamond in the shape of a square-based pyramid to produce an indentation, instead of a steel ball. All indentations are therefore geometrically similar and so there is no problem of achieving a ratio like the d/D ratio in the Brinell test. The angle between opposite faces of the diamond is 136°. This is the angle of indentation for a value of 0.375 for d/D in the Brinell test, of which the Vickers test is a development.

As with the Brinell test, the Vickers diamond hardness number (H_V value) is obtained by dividing the load in kg by the surface area of the indentation in mm^2. The surface area is given by $d^2/1.854$, where d is the length of the diagonal of the square of the indentation. The H_V value is therefore given by $1.854F/d^2$, where F is the load in kg, and d is in mm. The H_V and H_B values are equal where the latter is based on a d/D value of 0.375.

The most usual loads used are 10, 30 and 50 kg. These are much lighter than most Brinell loads, the indentations are very small and so the Vickers diamond test can be carried out on very small specimens. The Vickers diamond test is suitable for very hard materials and unlike the H_B value, H_V values are proportional to hardness. The diagonal of the Vickers impression is much easier to measure than the diameter of the Brinell impression, although it is advised that both diagonals should be measured.

(d) THE ROCKWELL HARDNESS TEST

The Rockwell Hardness Testing machine indicates hardness values directly on a dial. Unlike the Brinell and Vickers tests the Rockwell hardness number is based on depth of impression. To cover a wide range of materials, two penetrators are used, a 1/16 in diameter steel ball for the softer materials and a specially-developed diamond 'Brale' penetrator for the harder materials. The Brale penetrator is a cone with an angle of 120° and a spherical tip of 0.2 mm radius. Standard loads are used: an initial 10 kg in both cases with additional loads of 90 kg for the ball and 140 kg for the Brale penetrator.

Separate scales on the machine automatically make use of the additional depth of penetration due to the additional load after the initial load has overcome any problem of surface irregularities: a B scale (in red) for the ball, a C scale (in black) for the Brale. The dials are calibrated so that a small penetration, meaning a very hard material, gives a high reading. The scale used must be quoted, e.g. H_R C 55, H_R B 40.

The big advantage of the Rockwell test over Brinell and Vickers tests is the speed with which it can be made. As it is also independent of surface condition it is well suited to production line testing. British Standards, however, require hardness numbers to be based on the surface area of any indentation.

(e) OTHER HARDNESS TESTS

The Brinell, Vickers and Rockwell tests are examples of static hardness tests. Another example is the **Firth Hardometer test,** which is very similar to the Vickers test.

Examples of dynamic hardness tests are those using the **Herbert Pendulum Hardness Tester** and the **Shore Scleroscope**. The former uses an arched rocker resting on a steel or diamond pivot; hardness can be indicated by the time taken for ten single swings or by the difference between an initial angular displacement and the first swing. The Scleroscope is a portable apparatus in which a diamond-tipped hammer falls on to the material under test. The height of the rebound gives the hardness number.

(f) OTHER NON-DESTRUCTIVE TESTS

Flaws inside a casting can be revealed by X-ray methods.
Surface flaws can be revealed by electro-magnetic methods and by those using ultra-violet light. The former are applicable only to ferrous metals but the latter can be used for other metals and for other materials such as plastics and ceramics.
(See *Problems 5 to 7*)

3 **Impact tests**

The British Standard impact test is the **Izod test** in which a notched cantilever test-piece is broken by a striker fixed to a pendulum weight (*Fig 3*). The Izod value is the energy required to break the test-piece, indicated by the angle of swing after the impact, compared with the angle from which it first descends.

The test pieces may be of 10 mm square section or 0.45 in diameter circular section. The notch for the square-section test-piece is a V-notch with an angle of 45°, a root radius of 0.25 mm and a total depth of 2 mm. Some test-pieces have three notches appropriately arranged so that three tests can be carried out on the same test-piece.

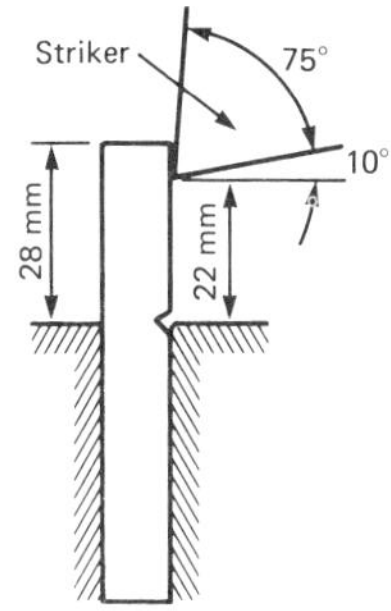

Fig 3

The Izod test is essentially an acceptance test – the result is either acceptable or not acceptable. It cannot be used to calculate the impact strength of the material under other conditions and there is no correlation with other tests such as tensile tests. The Izod test is often used to supplement a tensile test. It can indicate, as the tensile test cannot, whether or not effective heat treatment has been carried out on a steel. The radius at the root of the notch has a major effect on the values obtained.

In the **Charpy test** a beam-type test-piece is broken, as shown in plan in *Fig 4*. The test-piece shown is the Charpy test-piece but the machine is also used to test beam specimens with the Izod notch. The Charpy machine is calibrated to indicate impact resistance in kg m per cm² of area of fracture as well as the energy (kg m) absorbed in the fracture. Dividing the energy absorbed by area of fracture, however, does not give any better correlation with other tests than the simple energy value. The unit kg m, commonly used in these tests, is not a true energy unit. (See *Problems 8 and 9*)

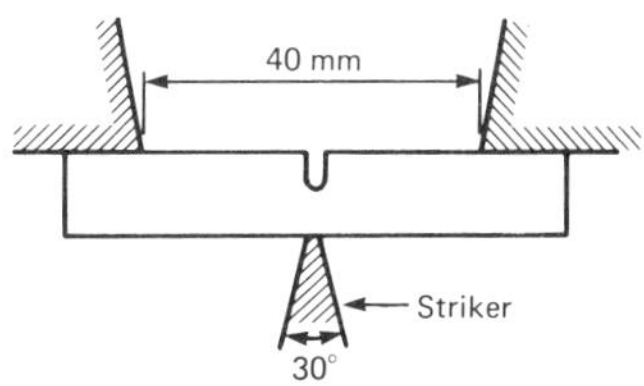

Fig 4

B. WORKED PROBLEMS ON MATERIALS

TEMPERATURE STRESS

Problem 1 A brass rod 210 mm long has its temperature raised through 30°C. If the coefficient of linear expansion of brass is 16×10^{-6} /°C, find the increase in length of the rod.

The increase in length is given by $l \alpha t$
where
l = original length;
α = coefficient of linear expansion;
t = temperature rise;

Hence, increase in length of bar $= 210 \text{ mm} \times 16 \times 10^{-6}/°\text{C} \times 30°\text{C}$
$= 210 \times 16 \times 30 \times 10^{-6}$ mm
$= 1008 \times 10^{-4}$ mm = **0.1008 mm**

Problem 2 A solid brass cylinder, 40 mm long, has its temperature reduced by 30°C below room temperature and is then inserted in a rigid clamp. Calculate the stress set up in the brass:
(a) if, at the reduced temperature, there is no clearance between cylinder and clamp,
(b) if at the reduced temperature there is a clearance of 0.01 mm between one end of the cylinder and the clamp.
Take E as 86 GPa and α as $16 \times 10^{-6}/°\text{C}$.

(a) The contraction due to stress must exactly balance the expansion due to temperature rise. That is,

$$\frac{\text{stress}}{E} \times l = l\,\alpha\,t$$

Thus stress $= E\,\alpha\,t = 86 \text{ GPa} \times 16 \times 10^{-6}/°\text{C} \times 30°\text{C}$
$= 86 \text{ GPa} \times 480 \times 10^{-6} = 86 \text{ MPa} \times 480 \times 10^{-3}$
= **41.28 MPa**

(b) The brass can expand by 0.01 mm without setting up any stress. The contraction due to stress will be 0.01 mm less than the expansion due to temperature rise. That is,

$$\frac{\text{stress}}{E} \times l = l\,\alpha\,t - 0.01 \text{ mm}$$

Hence, $\dfrac{\text{stress}}{86 \text{ GPa}} \times 40 \text{ mm} = 40 \text{ mm} \times 16 \times 10^{-6}/°\text{C} \times 30°\text{C} - 0.01 \text{ mm}$
$= 0.0192 \text{ mm} - 0.01 \text{ mm} = 0.0092 \text{ mm}$

from which, stress $= \dfrac{0.0092 \text{ mm} \times 86 \text{ GPa}}{40 \text{ mm}}$ = **19.78 MPa**

In both (a) and (b) the stress is compressive.

Problem 3 A steel bar 260 mm long is 20 mm in diameter for 140 mm of its length and 18 mm in diameter for the remaining 120 mm. It is cooled to 20°C below room temperature and then clamped between end restraints. If any expansion is prevented as the bar returns to room temperature, find the maximum stress set up in the bar. Take E as 200 GPa and α as $12.5 \times 10^{-6}/°\text{C}$.

The compressive force exerted by the end constraints is transmitted right through the bar and at every section is the product of stress and cross-sectional area. If the the stress is σ MPa where the diameter is 18 mm, then the force is,

$$\text{stress} \times \frac{\pi d^2}{4} = \sigma \text{ MPa} \times \frac{\pi(18 \text{ mm})^2}{4}$$

$$= \sigma \text{ MPa} \times 81\pi \text{ mm}^2 = 81\pi\sigma \text{ N}$$

Where the diameter is 20 mm the force is the same but the cross-sectional area is $\pi(10 \text{ mm})^2$. Therefore, the stress here is given by:

$$\frac{\text{force}}{\text{area}} = \frac{81\,\pi\sigma \text{ N}}{100\,\pi \text{ mm}^2} = \frac{81}{100}\,\sigma \text{ MPa}$$

The total contraction produced by compressive stresses is the sum of the contractions in the two lengths of 140 mm and 120 mm. Each contraction is

$$\text{strain} \times \text{length} = \frac{\text{stress}}{E} \times \text{length}$$

The total contraction due to stress is therefore:

$$\frac{\sigma \text{ MPa}}{200 \text{ GPa}} \times 120 \text{ mm} + \frac{0.81\sigma \text{ MPa}}{200 \text{ GPa}} \times 140 \text{ mm}$$

$$= \frac{\sigma \text{ MPa} \times 120 \text{ mm}}{200 \times 10^3 \text{ MPa}} + \frac{0.81\sigma \text{ MPa} \times 140 \text{ mm}}{200 \times 10^3 \text{ MPa}}$$

$$= \sigma \times 6 \times 10^{-4} \text{ mm} + 0.81\,\sigma \times 7 \times 10^{-4} \text{ mm}$$

$$= (6 + 5.67)\sigma \times 10^{-4} \text{ mm} = 11.67 \times 10^{-4}\,\sigma \text{ mm.}$$

This has to balance the expansion which would otherwise occur, due to the rise in temperature given by $l\,\alpha\,t = 260 \text{ mm} \times 12.5 \times 10^{-6}/°\text{C} \times 20°\text{C}$

$$= 0.065 \text{ mm}$$

i.e. $11.67 \times 10^{-4}\sigma = 0.065$

from which, $\sigma = \frac{650}{11.67}$ = **55.70 MPa.**

Problem 4 Two solid cylinders, each of 28 mm diameter, one of brass 40 mm long and the other of bronze 60 mm long, have their temperature reduced to 30°C below room temperature and then inserted end to end in a clamp, as shown in *Fig 5*.

The end clearance is measured as 0.004 mm at this temperature. Assuming the clamp to be perfectly rigid, what stresses will be set up when the cylinders return to room temperature? What forces will tend to open the jaws of the clamp? Take E as 86 GPa for brass and 100 GPa for bronze; take α as $16 \times 10^{-6}/°\text{C}$ for brass and $18 \times 10^{-6}/°\text{C}$ for bronze.

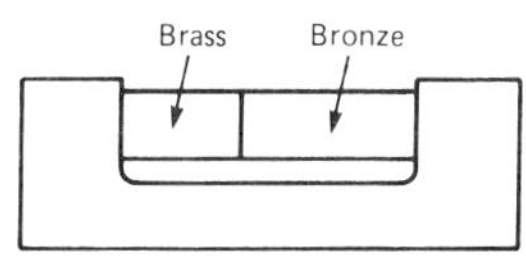

Fig 5

The **sum** of the expansions of the two cylinders due to temperature rise, and the **sum** of the contractions due to stress must be considered. The first will exceed the second by 0.004 mm.

The sum of the increases in length for 30°C rise (sum of products $l\,\alpha\,t$)
$= 40 \text{ mm} \times 16 \times 10^{-6}/°\text{C} \times 30°\text{C} + 60 \text{ mm} \times 18 \times 10^{-6}/°\text{C} \times 30°\text{C}$
$= 1.92 \times 10^{-2} \text{ mm} + 3.24 \times 10^{-2} \text{ mm} = 0.0516 \text{ mm.}$

The stress in the cylinders must therefore reduce their total length by 0.0516 mm − 0.004 mm = 0.0476 mm.

The stress will be the same in the two cylinders as they have the same cross-sectional area and transmit the same force. If this stress is σ MPa, then the sum of the contractions of the cylinders (sum of products, $\frac{\text{stress}}{E} \times$ length) is

$$\frac{\sigma \text{ MPa}}{86 \text{ GPa}} \times 40 \text{ mm} + \frac{\sigma \text{ MPa}}{100 \text{ GPa}} \times 60 \text{ mm}$$

$$= 4.6512\sigma \times 10^{-4} \text{ mm} + 6\sigma \times 10^{-4} \text{ mm} = 10.6512\sigma \times 10^{-4} \text{ mm.}$$

Equating this to the calculated contraction of 0.0476 mm gives:

$10.6512\sigma = 476$

thus $\sigma = \frac{476}{10.6512}$ = **44.69 MPa**

The forces on the clamp (stress × area)
$= 44.69$ MPa $\times \pi$ (14 mm)^2 = 27 518 N = **27.52 kN**

Further problems on temperature stress may be found in section C(b), Problems 1 to 12, page 11.

HARDNESS TESTS

Problem 5 In a Brinell hardness test a load of 3000 kg and a ball of 10 mm diameter were used. The diameter of the indentation was measured as 4.652 mm. Calculate the Brinell hardness number, the formula for spherical area being

$$\frac{\pi D}{2}\left[D - \sqrt{(D^2 - d^2)}\right]$$

where
D = diameter of ball,
d = diameter of impression.
Would this calculated value of Brinell hardness number be considered an acceptable one?

The Brinell hardness number is the load in kg divided by the spherical area of the indentation in mm^2.
From the formula, the spherical area of the indentation is

$$\frac{\pi \times 10}{2}\left[10 - \sqrt{(10^2 - 4.652^2)}\right] \text{ mm}^2$$

$= 5\pi\,(10 - 8.852)$ mm^2 = 18.032 mm^2.

The Brinell hardness number is therefore $\frac{3000}{18.032} =$ **166**

The ratio $\frac{d}{D}$ is $\frac{4.652 \text{ mm}}{10 \text{ mm}}$, i.e. 0.4652.

It is therefore within the acceptable range 0.25–0.50. It is not, however, close to the ideal value of 0.375.

The test might be repeated to give a value of d/D closer to 0.375, but this would require a non-standard load such as 2000 kg.

Problem 6 A Brinell hardness test is to be carried out on a piece of steel 2 mm thick. Balls of diameter 1 mm, 2 mm, 5 mm and 10 mm are available. The appropriate standard value of F/D^2 is to be used. The hardness number is expected to be about 400.

(a) Determine the largest acceptable ball, if the depth of impression may be taken as $\frac{F}{\pi DH}$
where, F = load in kg;
D = diameter of ball in mm;
H = Brinell hardness number.
(b) Determine the corresponding load to be applied.
(c) If the spherical area for the ball and load chosen is 1.829 mm^2, determine the actual Brinell hardness number.

(a) The depth of impression given by $\frac{F}{\pi DH}$ must not be greater than $t/10$, where t is the thickness of the material.

The estimated depth of impression is $\frac{F}{\pi D \times 400} = 0.000\,795\,77 \frac{F}{D}$

The limiting value for this is 0.2 mm, i.e., $0.000\,795\,77 \frac{F}{D} = 0.2$,

$$\text{giving } F/D = \frac{0.2}{0.000\,795\,77} = 251.33$$

Also, F/D^2 for steels has the standard value of 30,

Hence, $D = \frac{F/D}{F/D^2} = \frac{251.33}{30} = 8.378$ mm

A diameter of ball less than this must be chosen, i.e. **5 mm**

For a given value of F/D^2 the depth of impression is proportional to the ball diameter. A ball of diameter larger than the theoretical value of 8.378 mm would therefore give a depth of impression greater than the acceptable value of 0.2 mm.

(b) For $F/D^2 = 30$ and $D = 5$ mm, $F = 30 \times 25 =$ **750 kg.**

(c) Brinell hardness number $= \frac{\text{load (kg)}}{\text{spherical area of impression (mm}^2\text{)}}$

$$= \frac{750}{1.829} = \mathbf{410}$$

Problem 7 (a) In a Vickers diamond pyramid test a standard diamond was used under a load of 50 kg. The length of the diagonal d of the impression was measured as 0.537 mm. Taking the surface area of the impression as $d^2/1.854$ where d is the length of the diagonal of the square of the indentation, calculate the Vickers hardness number.

(b) If the same material were to be tested using a steel ball of 5 mm diameter, what diameter of impression would give the same hardness number as the Vickers number?

(a) The Vickers hardness number is given by $\frac{\text{load in kg}}{\text{surface area of impression in mm}^2}$

$$= \frac{50}{(0.537)^2/1.854} = \mathbf{321}$$

(b) The Vickers diamond pyramid test was designed to reproduce the ideal value of 0.375 for the Brinell ratio of d/D. This, then, is the ratio to give the same hardness number as that given in the Vickers test.

Diameter of impression = 5 mm × 0.375 = **1.875 mm**

Further problems on hardness tests may be found in section C(b), Problems 13 to 22, page 12.

IMPACT TESTS

Problem 8 The values obtained from Izod tests on two apparently identical steel specimens were:

Specimen A : 81.7 ft lb (11.32 kg m)

Specimen B : 14.7 ft lb (2.04 kg m)

What would you expect the ratio of the strengths of the steel to be (a) under other impact conditions, and (b) under tension?

(a) Specimen A will be stronger than specimen B under impact generally. The impact strengths of the steels, however, may be affected to different extends by the presence of the notch; it would be unreasonable to expect specimen A to be five times as strong as specimen B under all forms of impact loading.
(b) No conclusions at all can be drawn about the relative strengths of the steels in tension. These strengths could well be very similar.

Problem 9 An impact testing machine of the Izod type is required to give a maximum impact velocity of 4.5 m/s. if the effective length of the pendulum is 1.35 m, through what angle should the pendulum be raised before release? If the maximum energy in the pendulum is to be 2 kJ, what mass is required for the pendulum weight?

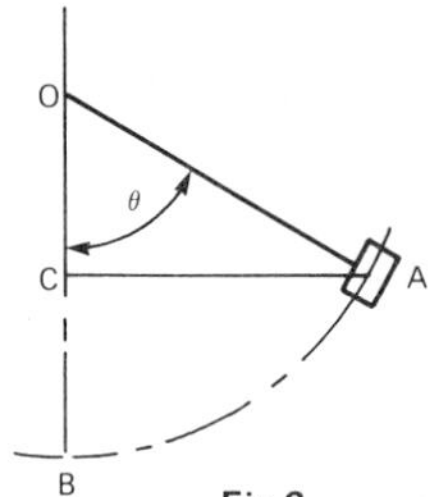

Fig 6

The vertical distance through which the pendulum weight must fall is given by $h = v^2/2g$, where v is the velocity at the bottom of the swing.

Thus, $h = \dfrac{(4.5\ \text{m/s})^2}{2 \times 9.81\ \text{m/s}^2} = 1.0321$ m.

This is the distance CB in *Fig 6*.
But CB = OB – OC = OA – OA cos θ.
The angle of swing θ required is therefore given by 1.0321 m = 1.35 m (1 – cos θ),

from which, $\cos\theta = 1 - \dfrac{1.0321}{1.35} = 0.2355$ and θ = **76.4°**.

The potential energy available for conversion to kinetic energy before impact is $Wh = Mgh$, where M is the mass of the pendulum weight,
i.e. $M \times 9.81\ \text{m/s}^2 \times 1.0321$ m = 2 kJ

Hence, $M = \dfrac{2000\ \text{Nm}}{9.81\ \text{m/s}^2 \times 1.0321\ \text{m}}$ = **197.5 kg**

Further problems on impact tests may be found in section C(b), Problems 23 to 26, page 13.

C. FURTHER PROBLEMS ON MATERIALS

(a) SHORT ANSWER PROBLEMS

1 What quantities are required to calculate the free expansion or contraction due to a change in temperature?

2 Define 'hardness'.

3 Briefly describe a Brinell hardness test.

4 State three disadvantages of the Brinell test compared with the Vickers pyramid diamond hardness test.

5 Briefly describe a Vickers hardness test.

6 Briefly describe the general principle of the Rockwell test.

7 Name two hardness tests other than Brinell, Vickers and Rockwell.

8 Describe an Izod test.

9 Describe a Charpy test.

10 Why is non-destructive testing needed?

(b) CONVENTIONAL PROBLEMS

Temperature stress

1 A steel rod is 324 mm long at 20°C. By how much will its length increase if its temperature rises to 85°C and it is free from any constraint? Take α as $12 \times 10^{-6}/°C$.

[0.2527 mm]

2 The length of a brass rod 176 mm long is found to increase by 0.16 mm when its temperature is raised through 56°C. What is the coefficient of linear expansion of brass, calculated from these values?

[$16.23 \times 10^{-6}/°C$]

3 A brass rod of 12 mm diameter is heated to 20°C above room temperature and its ends firmly clamped to prevent any further change in length. What stress will be set up in the rod when it returns to room temperature? What will be the forces on the ends of the rod? Take E as 90 GPa and α as $16 \times 10^{-6}/°C$.

[28.8 MPa; 3.257 kN]

4 A steel rod has its ends firmly clamped, completely preventing any change in length. What temperature change will induce a temperature stress in the rod of 40 MPa? Take E as 200 GPa and α as $12 \times 10^{-6}/°C$.

[16.67°C]

5 A cylinder of bronze 30 mm long is cooled to 30° below room temperature and then placed between the jaws of a clamp, the end clearance at this temperature being 0.004 mm. Assuming the clamp to be perfectly rigid, find the stress in the bronze when it returns to room temperature. Take E as 100 GPa and α as $18 \times 10^{-6}/°C$.

[40.67 MPa]

6 A steel rod, 42 mm long, is placed between the jaws of a clamp, the end clearance being 0.012 mm. What is the maximum permissible temperature rise in the rod if the stress in it is not to exceed 54 MPa? Take E as 200 GPa and α as $12 \times 10^{-6}/°C$ and assume that the jaws of the clamp remain a fixed distance apart.

[46.31°C]

7 A brass rod 128 mm long is placed between the jaws of a steel clamp, leaving an end clearance of 0.008 mm. The temperature of rod and clamp is raised though 42°C. Calculate the stress in the brass rod assuming that the jaws of the clamp move apart a distance of 0.0645 mm. Take E for brass as 90 GPa and α as $16 \times 10^{-6}/°C$.

[9.503 MPa]

8 An aluminium bar is heated through 28°C. Because of restraints at its ends its expansion is restricted to 70% of what it would be if unrestrained. What is the stress in the aluminium due to this temperature rise? Take E as 70 GPa and α as $24 \times 10^{-6}/°C$.

[14.11 MPa]

9 A brass bar is 18 mm in diameter for half of its length and 22 mm in diameter for the other half. It is heated to 35°C above room temperature and then its ends are

clamped, preventing any subsequent contraction. What will be the maximum stress in the bar when it returns to room temperature? Take E as 90 GPa and α as $16 \times 10^{-6}/°C$.

[60.38 MPa]

10 A steel bar 160 mm long is 15 mm in diameter for 40 mm of its length, 20 mm in diameter for 80 mm and 18 mm in diameter for the remaining 40 mm. If it is heated to 15°C above room temperature and its ends are then clamped to prevent any subsequent contraction, what will be the maximum stress set up in the bar when it returns to room temperature? Take E as 200 GPa and α as $12 \times 10^{-6}/°C$.

[51.07 MPa]

11 Two solid cylinders of the same diameter, one of brass 48 mm long and the other of aluminium 56 mm long, have their temperature reduced to 24°C below room temperature and are then inserted end to end in a clamp which prevents any subsequent change in their total length. What will be the stress in the cylinders when they return to room temperature? Take E as 90 GPa for brass and 70 GPa for aluminium. Take α as $16 \times 10^{-6}/°C$ for brass and $24 \times 10^{-6}/°C$ for aluminium.

[25.10 MPa]

12 Two solid cylinders of the same diameter, one of brass 58 mm long and the other of steel 36 mm long, are cooled to 42°C below room temperature and then inserted end to end in a clamp. The end clearance is measured as 0.01 mm at this temperature. Assuming that the jaws of the clamp remain at the same distance apart, what will be the stress in the cylinders when they return to room temperature? How much of the total expansion of 0.01 mm will be taken up by the steel cylinder? Take E as 90 GPa for brass and 200 GPa for steel. Take α as $16 \times 10^{-6}/°C$ for brass and $12 \times 10^{-6}/°C$ for steel.

[57.15 MPa; 0.00786 mm]

Hardness tests

13 In a Brinell hardness test a load of 3000 kg and a ball of 10 mm diameter were used. The diameter of the indentation was 3.85 mm. Calculate the Brinell hardness number, using the formula for spherical area: $\frac{\pi D}{2}\left[D - \sqrt{(D^2 - d^2)}\right]$

where D = diameter of ball,
d = diameter of impression.

[248]

14 Had it been necessary for the hardness of the material of problem 13 to be tested using a 5 mm ball, what load would have been appropriate? What diameter of impression would then have been obtained?

[750 kg; 1.925 mm]

15 A series of Brinell hardness tests was carried out on an unidentified material, using a 10 mm ball. The loads used and the corresponding diameters of impression were as follows:

3000 kg	2000 kg	1500 kg	500 kg
4.65 mm	3.80 mm	2.81 mm	2.06 mm

Determine the most reliable Brinell hardness number available from these tests.

[169, under 2000 kg load]

16 A Brinell hardness test is to be carried out on a piece of brass 1.5 mm thick. Balls of diameter 1 mm, 2 mm, 5 mm and 10 mm are available. The appropriate standard value of F/D^2 is to be used. The hardness number is expected to be about 90.

Determine the largest diameter of ball acceptable and the corresponding load to be applied. The depth of impression may be taken as

$$\frac{F}{\pi DH}$$

where
F = load in kg;
D = diameter of ball in mm;
H = Brinell hardness number.

[2 mm; 40 kg]

17 A Brinell hardness test is to be carried out on a steel bar of 10 mm square section. Balls of diameter 1 mm, 2 mm, 5 mm and 10 mm are available. Determine the largest diameter of ball acceptable for the test and, using the appropriate standard value of F/D^2, the load to be applied. If the actual ratio of d/D is found to be 0.395, calculate the Brinell hardness number.

[5 mm; 750 kg; 235]

18 The Brinell hardness number of a specimen of steel is 243. Use a correlation factor to estimate the ultimate tensile stress of the steel. How accurate would you expect this estimate to be?

[730 MPa to 850 MPa]

19 In a Vickers Diamond Pyramid test a standard diamond was used under a load of 120 kg. The length of the diagonal of the impression was measured as 0.642 mm. Calculate the Vickers hardness number.

[540]

20 Sketch and describe the Brale penetrator of a Rockwell hardness testing machine. State the main advantage and the main disadvantage of the Rockwell test compared with the Vickers test.

21 Name two hardness tests other than Brinell, Vickers and Rockwell tests. Explain briefly how they are carried out.

22 Explain briefly why non-destructive testing is often necessary. Give two examples of non-destructive tests, one for surface flaws and one for internal flaws.

Impact tests

23 Sketch and describe a standard Izod test-piece, showing the details of the notch and how the test-piece is positioned for a test.

24 A standard Izod test-piece of square section absorbs 12.4 kg m of energy in the test. Express the results in terms of:
(a) kg m/cm^2 of area of fracture,
(b) kg m/cm^3 of volume of test-piece above the clamp.
Use the standard test-piece dimensions.
Which of the three figures for the test gives an indication of tensile strength?

[15.5 kg m/cm^2; 4.43 kg m/cm^3; none]

25 An impact testing machine of the Izod type is required to give an impact velocity of 4 m/s. If the effective length of the pendulum is 1.25 m, through what angle should the pendulum be raised before release? If the mass of the pendulum is 180 kg, what energy does it possess immediately before impact: give the answer in kg m and in kJ.

[69.7°, 146.8 kg m = 1.44 kJ]

26 Describe briefly the main differences between an Izod test and a Charpy test. Which of these tests give the better correlation with (a) tensile strength, (b) hardness?

2 Dynamics

A. MAIN POINTS CONCERNING TORQUE AND THE KINETIC ENERGY OF ROTATING BODIES

1 **Kinetic energy of rotation**

The tangential velocity v and angular velocity ω (*Fig 1*) are connected by the formula $v = \omega r$.

(v is in metres per second if ω is in radians per second and r in metres).

The kinetic energy of a particle of mass m, is

$$\frac{mv^2}{2} = \frac{m(\omega r)^2}{2} = \frac{m\omega^2 r^2}{2}$$

If m is in kg and v in m/s then KE will be in N m or joules (J):
kg $(m/s)^2$ can be written kg $m/s^2 \times m$ and kg m/s^2 = N,
Hence, kg $(m/s)^2$ = N m = J
The joule is the basic unit of most forms of energy.

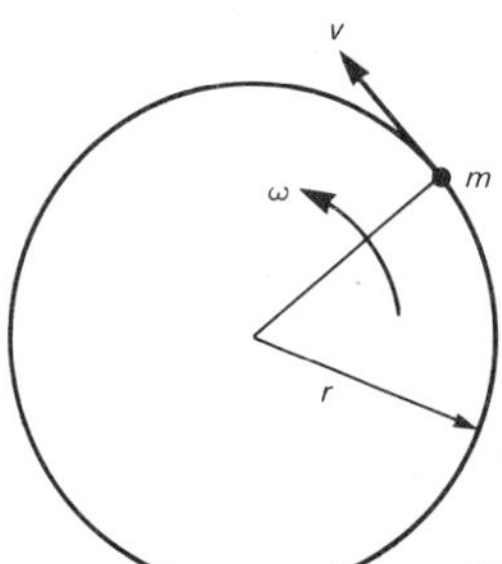

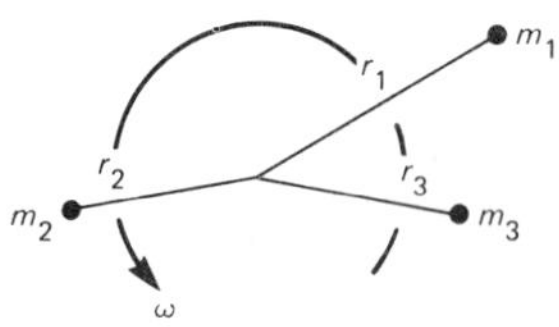

Fig 2

Fig 1

The total kinetic energy of a system of masses rotating at different radii but with the same angular speed ω as in *Fig 2* is:

$$\frac{m_1 \omega^2 r_1^2}{2} + \frac{m_2 \omega^2 r_2^2}{2} + \frac{m_3 \omega^2 r_3^2}{2}$$

$$= (m_1 r_1^2 + m_2 r_2^2 + m_3 r_3^2) \frac{\omega^2}{2}$$

In general terms this can be written –

$$\Sigma \frac{m\omega^2 r^2}{2} = (\Sigma m r^2) \frac{\omega^2}{2} = \frac{I\omega^2}{2} = \frac{M k^2 \omega^2}{2}$$

where I (= Σmr^2) is called the **moment of inertia** of the system about the axis of rotation. The system may well be a solid body.

M (= Σm) is the total mass. k is called the **radius of gyration** of the system for the given axis. If all of the mass were concentrated at the radius of gyration it would give the same moment of inertia as the actual system.
(See *Problems 1 and 2*)

2 **Torque and angular acceleration**

$T = I\alpha$,

where, T is the accelerating torque;
I is the moment of inertia;
α is the angular acceleration.

If I is in kg m^2 and α in rad/s^2, T will be in N m.
The formula derives from Newton's second law of motion.
If a torque T (N m) is exerted through an angle θ (radians) the work done is given by $T\theta$ (N m or J).
If this work is available to increase the kinetic energy of a rotating body of moment of inertia I, then

$$T\theta = \frac{I}{2}(\omega_2^2 - \omega_1^2)$$

where ω_1, ω_2 are the initial and final angular velocities,

i.e. $T\theta = I\left(\frac{\omega_2 + \omega_1}{2}\right)(\omega_2 - \omega_1)$

But $\frac{\omega_2 + \omega_1}{2}$ is the mean angular speed $= \frac{\theta}{t}$ where t is the length of time involved.

Also $(\omega_2 - \omega_1)$ is the change in angular speed, i.e. $\alpha\, t$.

Hence, $T\theta = I\left(\frac{\theta}{t}\right)(\alpha\, t)$ and $T = I\alpha$

This shows that the energy method is an alternative to a more direct application of Newton's second law.
(See *Problems 3 and 4*)

3 **Flywheels**

The function of a flywheel is to restrict fluctuations of speed by absorbing and releasing large quantities of kinetic energy for small speed variations. To do this they require large moments of inertia and to avoid excessive mass they need to have radii of gyration as large as possible. Most of the mass of a flywheel is usually in its rim. (See *Problem 5*)

4 **Belt drives**

(i) The ratio of the tensions for a flat belt **when the belt is on the point of slipping** is $\frac{T_1}{T_2} = e^{\mu\theta}$
where
μ is the coefficient of friction between belt and pulley,
θ is the angle of lap, in radians, (see *Fig 3*), and
e is the exponent 2.718 (see chapter 8).

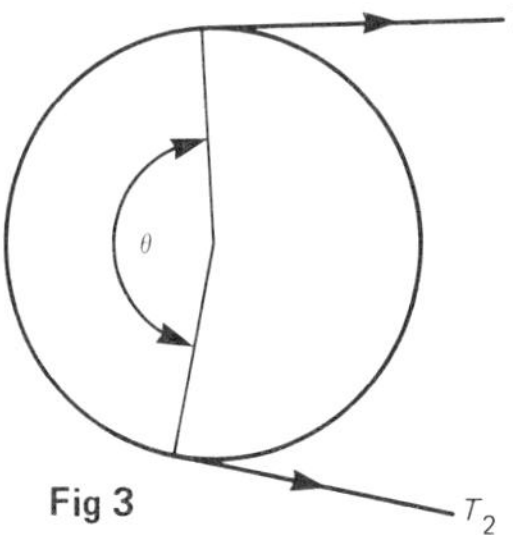

Fig 3

For a vee belt as in *Fig 4* the ratio is $\frac{T_1}{T_2} = e^{\frac{\mu\theta}{\sin\alpha}}$,

where α is the half angle of the groove and of the belt.

This gives a much larger ratio than for the flat belt. The V-belt is jammed into its groove and is less likely to slip. Referring to *Fig 4*, the force of friction on each side is μR_N where R_N is the normal (perpendicular) reaction on each side.

The triangle of forces shows that

$R_N = \frac{R/2}{\sin\alpha}$ where R is the resultant reaction.

The force of friction giving rise to the difference between T_1 and T_2 is therefore

$\mu R_N \times 2 = \frac{\mu}{\sin\alpha} R$

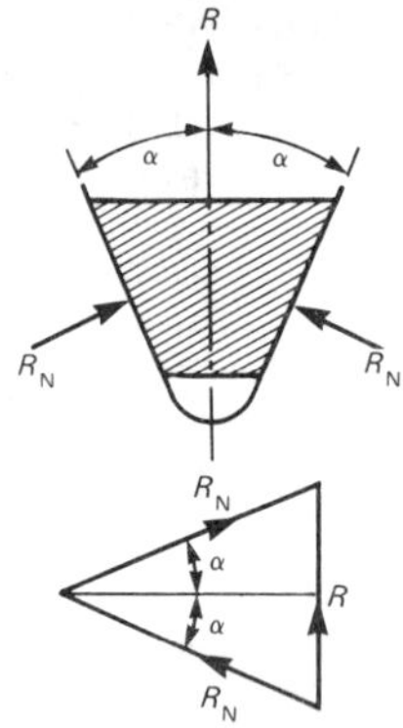

Fig 4

The corresponding force of friction for a flat belt is μR. Comparing the forces of friction for flat and V-belts it can be said that the V-belt is equivalent to a flat belt with a coefficient of friction given by $\frac{\mu}{\sin\alpha}$.

(ii) The net force exerted on the belt by a driving pulley or by the belt on a driven pulley is $T_1 - T_2$. The power transmitted by a belt is therefore $(T_1 - T_2)\,v$, where v is the speed of the belt. If $(T_1 - T_2)$ is measured in newtons and v in metres per second the power will be in N m/s, that is, watts (W).

The belt speed is given in m/s by ωr, where ω is the angular speed of the pulley (rad/s) and r is its radius (m). Also

$\omega = \frac{2\pi N}{60}$ where N is the pulley speed in rev/min.

It should be noted that:

(a) slipping would be expected to occur first on the pulley having the smaller angle of lap;

(b) in practice care would be taken to ensure that slipping is not likely to occur.

(See *Problems 6 and 7*)

B. WORKED PROBLEMS ON DYNAMICS

KINETIC ENERGY OF ROTATION

Problem 1. A small body, of mass 82 g, is fastened to a wheel and rotates in a circular path of 456 mm diameter. Calculate the increase in kinetic energy of the body when the speed of the wheel increases from 450 rev/min to 950 rev/min.

The increase in kinetic energy is given by: $\frac{I(\omega_2{}^2 - \omega_1{}^2)}{2}$, where

I is the moment of inertia ($= Mk^2$),
ω_2 is the final angular speed;
ω_1 is the initial angular speed.

The kinetic energy will be given in J if I is in kg m^2 and ω_2 and ω_1 in rad/s.

Here $I = 0.085 \text{ kg} \times (\frac{0.456 \text{ m}}{2})^2$

$= 4.2627 \times 10^{-3}$ kg m^2

$$\omega_2 = \frac{950 \times 2\pi}{60} = 99.484 \text{ rad/s}$$

$$\omega_1 = \frac{450 \times 2\pi}{60} = 47.124 \text{ rad/s}$$

Therefore the increase in KE

$$= \frac{4.2627 \times 10^{-3} (99.484^2 - 47.124^2)}{2} \text{ J}$$

= **16.36 J.**

Problem 2 A system consists of three small masses rotating at the same speed about the same fixed axis. The masses and their radii of rotation are: 16 g at 256 mm, 23 g at 192 mm and 31 g at 176 mm.

(a) Find the moment of inertia of the system about the given axis and the corresponding radius of gyration.

(b) If the speed of rotation is 1250 rev/min, find the kinetic energy in the system.

(a) It is assumed that the dimensions of the bodies are negligible compared with the radii of rotation. In these circumstances the radius of gyration of each body is the same as its radius of rotation. The moment of inertia of the system about the given axis is then:

$\Sigma mr^2 = 16 \text{ g} \times (0.256 \text{ m})^2 + 23 \text{ g} \times (0.192 \text{ m})^2 + 31 \text{ g} \times (0.176 \text{ m})^2$

$= 1.0486 \text{ g m}^2 + 0.8479 \text{ g m}^2 + 0.9603 \text{ g m}^2$

$=$ **2.857 g m^2** or **2.857 kg mm^2**

This is also $M k^2$ where M is the total mass and k is the overall radius of gyration.

$M = 16 \text{ g} + 23 \text{ g} + 31 \text{ g} = 70 \text{ g}$

$70 \text{ g} \times k^2 = 2.857 \text{ g m}^2$,

from which, $k = \sqrt{\left(\frac{2.857}{70}\right)}$ m = 0.2020 m = **202.0 mm**

(b) The kinetic energy in each mass could be calculated and the values added together. As the moment of inertia of the system is known, however, its kinetic energy can be obtained from $\frac{I\omega^2}{2}$ where I is the moment of inertia of the system. That is, the kinetic energy is

$$\frac{2.857 \times 10^{-3} \text{ kg m}^2}{2} \times (\frac{1250 \times 2\pi}{60} \text{ rad/s})^2 = \textbf{24.48 J.}$$

Further problems on kinetic energy of rotation may be found in section C(b), Problems 1 to 9, page 22.

TORQUE AND ANGULAR ACCELERATION

Problem 3 A shaft with its rotating parts has a moment of inertia of 16.42 kg m^2. It is accelerated from rest by an accelerating torque of 43.6 N m. Find the speed of the shaft (a) after 15 s, and (b) after the first 4 revolutions.

(a) The angular acceleration α is given by $T = I\alpha$,
where
T is the accelerating torque,
I is the moment of inertia.

Thus, $\alpha = \frac{T}{I} = \frac{43.6 \text{ N m}}{16.42 \text{ kg m}^2} = 2.6553 \text{ rad/s}^2$.

The angular speed of the shaft is initially zero and therefore after 15 s the angular acceleration X time

i.e. $2.6553 \text{ rad/s}^2 \times 15 \text{ s} =$ **39.83 rad/s**

$= 39.83 \times \frac{2\pi}{60}$ rev/min $=$ **380.3 rev/min**

(b) The work done by the accelerating torque T (43.6 N m) through θ (4 revolutions) is $T\theta = 43.6 \text{ N m} \times (4 \times 2\pi) \text{ rad}$
$= 1095.8 \text{ N m} = 1.0958 \text{ kJ}$

This appears as kinetic energy, given by $\frac{I\omega^2}{2}$ where I is the moment of inertia and ω is the angular speed acquired from rest.

i.e. $1095.8 \text{ J} = \frac{16.42 \text{ kg m}^2 \times \omega^2}{2}$

from which, $\omega = \sqrt{\left(\frac{1095.8 \times 2}{16.42}\right)}$ rad/s

$= 11.55$ rad/s $=$ **110.3 rev/min**

It should be noted that the application of $T = I\alpha$, which derives from Newton's second law of motion, and equating work done to energy gained are alternative approaches always applicable in problems of this type.

Where time is specified, as in part (a), the application of $T = I\alpha$ has some advantage but the other method can be used. Taking ω as the angular speed (rad/s) after 15 seconds, the average speed is given by $\frac{\omega}{2}$ and the angular movement as $\frac{\omega}{2} \times 15$ (rad). Then the product $T\theta$ can be equated to the kinetic energy gained, given by $\frac{I\omega^2}{2}$.

That is, $T\theta = T \times \frac{\omega}{2} \times 15 = \frac{I\omega^2}{2}$

The angular speed ω, being the only unknown, may be obtained from this equation. Where the movement is specified, as in part (b), the energy method has some advantage, but again either method can be used.

Problem 4 The driving torque on a turbine rotor is 230 N m; the sum of frictional and other resisting torques is 27 N m.
(a) What is the gain in kinetic energy of the rotor while it turns through 100 revolutions?
(b) If the moment of inertia of the rotor is 23.2 kg m^2 and the speed at the beginning of the 100 revolutions is 600 rev/min, what will be its speed at the end?

(a) The net torque accelerating the rotor and so increasing its kinetic energy is 230 N m − 27 N m = 203 N m.

The kinetic energy gained is equal to the work done by this torque over 100 revolutions. This is given by

$T\theta$ = 203 N m × (100 × 2π) radians = **127.55 kJ.**

(b) The initial kinetic energy of rotation is given by:

$$\frac{I\,\omega_1{}^2}{2}\text{ ; where } \omega_1\text{, the original angular velocity}$$

$$= \frac{600 \times 2\pi}{60}\ \text{rad/s} = 62.832\ \text{rad/s}$$

$$\frac{I\omega_1{}^2}{2} = \frac{23.2\ \text{kg m}^2 \times (62.832\ \text{rad/s})^2}{2} = 45.80\ \text{kJ}$$

The final KE is the sum of initial KE and KE gained,

$$\frac{I\omega_2{}^2}{2} = 45.80\ \text{kJ} + 127.55\ \text{kJ} = 173.35\ \text{kJ}$$

Therefore, the final angular velocity,

$$\omega_2 = \sqrt{\left(\frac{17\,350\ \text{N m} \times 2}{23.2\ \text{kg m}^2}\right)} = 122.25\ \text{rad/s}$$

$$= \frac{122.25 \times 60}{2\pi}\ \text{rev/min} = \mathbf{1167\ rev/min}$$

Further problems on torque and angular acceleration may be found in section C(b), Problems 10 to 15, page 22.

FLYWHEELS

Problem 5 A cast iron flywheel is required to release 2.10 kJ of kinetic energy when its speed falls from 3020 rev/min to 3010 rev/min. The moment of inertia of the flywheel may be assumed to be concentrated in its rim which is to be of rectangular section, the external and internal diameters being 670 mm and 600 mm. The radius of gyration of the rim may be assumed to be its mean radius. Taking the density of cast iron as 7.8 tonne/m³ (7800 kg/m³) calculate the width of rim required for the flywheel.

The change of KE is given by $\dfrac{I(\omega_2{}^2 - \omega_1{}^2)}{2}$. Hence the moment of inertia I required for the flywheel is given by: $\dfrac{\text{change of KE} \times 2}{\omega_2{}^2 - \omega_1{}^2}$

$$= \frac{2100\ \text{J} \times 2}{\left(\frac{3020 \times 2\pi}{60}\ \text{rad/s}\right)^2 - \left(\frac{3010 \times 2\pi}{60}\ \text{rad/s}\right)^2} = 6.3515\ \text{kg m}^2$$

This is also given by $M\,k^2$ where M is the mass of the rim and k is the radius of gyration, assumed here to be the mean radius of the rim:

$$\frac{\dfrac{670\ \text{mm}}{2} + \dfrac{600\ \text{mm}}{2}}{2} = 317.5\ \text{mm}$$

Hence, $M = \dfrac{I}{k^2} = \dfrac{6.3515\ \text{kg m}^2}{(0.3175\ \text{m})^2} = 63.007\ \text{kg}$

This is the product of the volume V and density ρ of the cast iron rim.

$$V = \frac{M}{\rho} = \frac{63.007\ \text{kg}}{7800\ \text{kg/m}^3} = 0.008\,077\,8\ \text{m}^3$$

The volume V is given by $\frac{\pi b}{4}(D^2 - d^2)$

where
b is the width of the rim,
D, d are the external and internal diameters, respectively

Hence, $$b = \frac{4V}{\pi(D^2 - d^2)} = \frac{4 \times 0.008\,077\,8\ \text{m}^3}{\pi(0.67^2 - 0.6^2)\ \text{m}^2}$$
$$= 0.1157\ \text{m} = \mathbf{115.7\ mm}$$

Further problems on flywheels may be found in section C(b), Problems 16 to 21, page 23.

BELT DRIVES

Problem 6 A pulley has vee grooves to take six belts which are required to transmit 125 kW from the pulley when its speed is 1920 rev/min. The angle of the grooves is 45°, the angle of lap is 175° and the effective diameter of the pulley is 265 mm. The coefficient of friction may be taken as 0.32. Calculate the lowest possible value for the tension on the tight side of each belt.

The power transmitted by the six belts is given by $6(T_1 - T_2)v$ where T_1, T_2 are the tensions on tight and slack sides respectively of one belt,
and v is the belt speed = ωr.

i.e. $$125\,000\ \text{W} = 6(T_1 - T_2)\left(1920 \times \frac{2\pi}{60}\ \text{rad/s}\right)\left(\frac{0.265\ \text{m}}{2}\right)$$

$$T_1 - T_2 = \frac{125\,000\ \text{W}}{6 \times \frac{1920 \times 2\pi}{60} \times 0.1325\ \text{m/s}} = 782.01\ \text{N}$$

The tension T_1 will have its lowest possible value when the belts are on the point of slipping. For this condition $\frac{T_1}{T_2} = e^{\left(\frac{\mu}{\sin\alpha}\right)\theta}$,

where
μ is the coefficient of friction
α is the half angle of the grooves
θ is the angle of lap in radians

$$\frac{T_1}{T_2} = e^{\left[\frac{0.32}{\sin 22.5^\circ}\left(\frac{175\pi}{180}\right)\right]} = 12.859$$

i.e. $T_1 = 12.859\,T_2$ or $T_2 = \frac{T_1}{12.859}$.

Hence, $$T_1 - T_2 = T_1\left(1 - \frac{1}{12.859}\right) = 0.92223\,T_1$$

The value of $T_1 - T_2$ has been calculated as 782.01 N.

The lowest possible value for T_1 is therefore $\frac{782.01\ \text{N}}{0.92223} = \mathbf{848.0\ N}$

Problem 7 A flat belt is required to transmit 25 kW from a pulley on which the angle of lap is 165°. The belt speed is 21.5 m/s, the coefficient of friction between belt and pulley is 0.28 and the maximum permissible tensile stress in the belt is 2.05 MPa (2.05 MN/m^2). If the width of the belt is to be ten times its thickness, calculate the minimum acceptable dimensions for the belt.

The power transmitted by the belt is given by $(T_1 - T_2)v$,
where T_1, T_2 are the tensions on the tight and slack side respectively, and v is the belt speed,
i.e. 25 000 W = $(T_1 - T_2)$ 21.5 m/s

$$T_1 - T_2 = \frac{25\,000 \text{ W}}{21.5 \text{ m/s}} = 1162.8 \text{ N}$$

The tension T_1 will have its lowest possible value when the belt is on the point of slipping. For this condition, $\frac{T_1}{T_2} = e^{\mu\theta}$,

where
μ is the coefficient of friction, and
θ is the angle of lap, in radians,

i.e. $\frac{T_1}{T_2} = e^{\frac{(0.28 \times 165\pi)}{180}} = 2.2397$

i.e. $T_1 = 2.2397\, T_2$ or $T_2 = \frac{T_1}{2.2397}$

Hence $T_1 - T_2 = T_1 \left(1 - \frac{1}{2.2397}\right) = 0.55351\, T_1$.

The value of $T_1 - T_2$ has been calculated as 1162.8 N. The least possible value of T_1 is therefore $\frac{1162.8 \text{ N}}{0.55351} = 2100.8$ N.

The maximum stress in the belt is given by $\frac{T_1}{\text{cross-sectional area}}$

$= \frac{T_1}{t \times 10t} = \frac{2100.8 \text{ N}}{10\, t^2}$ where t = thickness of belt.

This stress must not exceed 2.05 MN/m^2, which is equal to 2.05 N/mm^2.

That is, for minimum permissible dimensions, 2.05 N/mm^2 $= \frac{2100.8 \text{ N}}{10\, t^2}$

Hence, thickness of belt, $t = \sqrt{\left[\frac{2100.8 \text{ N}}{10 \times 2.05 \text{ N/mm}^2}\right]}$ = **10.12 mm**

and width of belt = 10 t = **101.2 mm.**

Further problems on belt drives may be found in section C(b), following, Problems 22 to 30, page 24.

C. FURTHER PROBLEMS ON DYNAMICS

(a) SHORT ANSWER PROBLEMS

1 If m is the mass of particle rotating at an angular velocity ω at a radius r its kinetic energy is given by . . . .

2 What is meant by the radius of gyration?

3 The kinetic energy of a body rotating at an angular speed ω and having a moment of inertia I is given by . . . .

4 If I is the moment of inertia and α the angular acceleration then the accelerating torque T is given by T = ...

5 Discuss the use of a rotating flywheel as a source of energy.

6 The tension T_1 of one side of a flat belt when the belt is on the point of slipping is given by $T_1 = T_2 e^{\mu\theta}$. Explain the symbols used in the formula.

(b) CONVENTIONAL PROBLEMS

Kinetic energy of rotation

1 A small body of mass 12 g moves at a uniform speed around a circle of diameter 254 mm, making 24 complete revolutions every second. What is the kinetic energy of the body?

[2.20 J]

2 A small body of mass 37 g moves in a circle of 218 mm diameter at the end of a rod. What will be the increase in kinetic energy of the body if the angular speed of the rod increases from 42 rad/s to 85 rad/s?

[1.20 J]

3 A small body of mass 112 g is fastened to the rim of a wheel of 980 mm diameter, rotating about a fixed central axis. Calculate the increase in kinetic energy of the body when the speed of the wheel increases from 620 rev/min to 1350 rev/min.

[212.0 J]

4 A light rod, 320 mm long, carries concentrated masses of 32 g and 45 g at its ends. Find the moment of inertia and radius of gyration with respect to an axis at right angles to the rod and 140 mm from the 45 g mass.

[1.919 g m^2 or 1919 kg mm^2; 157.9 mm]

5 If the light rod of problem 4 rotates at 1500 rev/min about the given axis, what is its kinetic energy? [23.67 J]

6 A system consists of three small masses rotating together about a fixed axis. The masses and their radii of rotation are: 124 g at 125 mm; 58 g at 132 mm; 76 g at 148 mm. Find their combined moment of inertia about the given axis and the corresponding radius of gyration.

[4.613 g m^2 or 4613 kg mm^2; 133.7 mm]

7 If the system of three forces of problem 6 rotates at 2100 rev/min, what is the kinetic energy in the system?

[111.5 J]

8 A disc of mass 2.75 kg has a radius of gyration of 126 mm. What is its kinetic energy at a speed of 2500 rev/min?

[1.496 kJ]

9 A steel disc of 450 mm diameter and uniform thickness of 28 mm rotates about its central axis at a speed of 3200 rev/min. Calculate its kinetic energy. Take the density of steel as 7.80 tonne/m^3.
(The radius of gyration of a disc of uniform thickness about its central axis is given by $\frac{\text{radius of disc}}{\sqrt{2}}$.) [49.37 kJ]

Torque and angular acceleration

10 A uniform torque of 2.76 kN m is applied to a shaft while it turns through 24 revolutions. Assuming no frictional or other resistances, calculate the increase in

kinetic energy of the shaft. If the shaft is initially at rest and its moment of inertia is 23.6 kg m^2, find its rotational speed, in revolutions per minute at the end of the 24 revolutions.

[416.2 kJ; 1793 rev/min]

11 An accelerating torque of 25 N m is applied to a rotor while it turns through 8 revolutions. What is its increase in kinetic energy? If the moment of inertia of the rotor is 11.6 kg m^2 and its speed at the beginning of the 8 revolutions is 1200 rev/min, what will be its speed at the end?

[1.257 kJ; 1208 rev/min]

12 A shaft with the rotating parts attached to it has a moment of inertia of 51.6 kg m^2. What uniform torque is required to accelerate the shaft from rest to a speed of 1600 rev/min while it turns through 16 revolutions?

[7.205 kN m]

13 A shaft with the rotating parts attached to it has a moment of inertia of 28.6 kg m^2. What angular acceleration of the shaft would be produced by an accelerating torque of 472 N m?

[16.50 rad/s^2]

14 A shaft with the rotating parts attached to it has a moment of inertia of 42.6 kg m^2. What torque is required to give it an angular acceleration of 4.2 rad/s^2

[178.9 N m]

15 A turbine rotor has a mass of 624 kg and a radius of gyration of 135 mm. What torque is required to accelerate it from rest to a speed of 3600 rev/min in 24 seconds?

[178.6 N m]

Flywheels

16 A flywheel has a mass of 267 kg and a radius of gyration of 156 mm. How much kinetic energy will it lose when it slows down from 1850 rev/min to 1820 rev/min?

[3.923 kJ]

17 A flywheel has a mass of 318 kg and a radius of gyration of 208 mm. It absorbs 11.2 kJ of kinetic energy, its speed increasing from an initial value of 1250 rev/min. What is its final speed?

[1308 rev/min]

18 A flywheel is required to absorb 72.4 kJ of energy while its speed increases from 2000 rev/min to 2050 rev/min. Calculate the moment of inertia required for the flywheel.

[65.20 kg m^2]

19 A flywheel is required to absorb 12.8 kJ of energy while its speed increases from 2320 rev/min to 2340 rev/min. If the maximum obtainable radius of gyration of the flywheel is 220 mm, find the minimum possible mass for the flywheel.

[517.5 kg]

20 The rim of a cast iron flywheel has an external diameter of 878 mm, a radial thickness of 30 mm and a width of 108 mm. Find the moment of inertia of the flywheel, assuming that 96% of this is in the rim, and assuming that the radius of gyration of the rim is its mean radius. Take the density of cast iron as 7.78 tonne/m^3.

[12.58 kg m^2]

21 Calculate the kinetic energy given up by the flywheel of problem 20 when its speed drops from 960 rev/min to 950 rev/min.

[2.634 kJ]

Belt drives

22 A flat belt has an angle of lap of 165° on a pulley; the coefficient of friction between the two is 0.32. If the tension on the tight side of the belt is 840 N what is the minimum tension on the other side if slipping is not to occur?

[334.2 N]

23 If the speed of the belt in problem 22 is 32 m/s, what is the maximum power which can be transmitted?

[14.90 kW]

24 A flat belt has an angle of lap of 170° on a pulley. The coefficient of friction between belt and pulley is 0.28. A difference of 720 N is required between the tensions on the tight and slack sides of the belt when it is transmitting power. What is the lowest possible value of the tension on the tight side of the belt?

[1.276 kN]

25 A pulley is grooved to take a vee-belt with a groove angle of 50° and an angle of lap of 175°. The tension on the tight side of the belt is 640 N and the coefficient of friction between belt and pulley is 0.35. Calculate the minimum permissible tension on the slack side of the belt if slipping is not to occur.

[51.0 N]

26 If the belt speed in *Problem 25* is 27 m/s, calculate the maximum power which can be transmitted.

[15.90 kW]

27 A pulley has vee grooves to take eight belts. The angle of the grooves is 40°, the angle of lap is 160° and the coefficient of friction is 0.3. If the belt speed is 22 m/s and the maximum permissible tension in each belt is 1.25 kN, calculate the maximum power which the belt can transmit.

[201 kW]

28 A belt drive is required to transmit 120 kW from a pulley rotating at 1950 rev/min. The pulley is to have vee grooves to take a number of belts, the angle of the grooves being 45°. The maximum permissible tension in each belt is 875 N and the angle of lap is 165°. Taking the coefficient of friction between belts and pulley as 0.32 and the effective diameter of the pulley as 218 mm, determine the minimum number of belts required.

[7]

29 A flat belt 75 mm wide and 5 mm thick transmits 12.5 kW from a pulley of 280 mm diameter rotating at 1900 rev/min. The angle of lap is 172° and the coefficient of friction between belt and pulley is 0.32. Calculate the maximum stress in the belt.

[1.938 MPa]

30 A flat belt 6.0 mm thick is required to transmit 18 kW from a pulley on which the angle of lap is 170°. The coefficient of friction between belt and pulley is 0.33, the belt speed is 24.5 m/s and the maximum permissible tensile stress in the belt is 2.14 MPa (2.14 MN/m^2). Calculate the minimum width of belt required.

[90.40 mm]

3 Statics and hydrostatics

A. MAIN POINTS CONCERNING STATICS AND HYDROSTATICS

1 **Static equilibrium of non-concurrent forces**
A body at rest is in a condition of static equilibrium. This requires that any forces acting on it are completely balanced, that is, the resultant force is zero and the resultant moment about any point is zero. These conditions can be expressed as:
$\Sigma H = 0$, i.e., the net horizontal component force is zero
$\Sigma V = 0$, i.e., the net vertical component force is zero
$\Sigma M = 0$, i.e., the net moment about any point is zero.
For three forces to be in equilibrium they must be concurrent or parallel. This restriction does not apply to systems of four or more forces.

In most problems involving four or more forces it is best to look at moments first, for by taking moments about the point of intersection of two unknown forces, these unknown forces can be ignored.
(See *Problems 1 to 7*)

HYDROSTATIC PRESSURE

2 Density (ρ) is mass per unit volume and its basic unit is kg/m^3. The density of fresh water is approximately 1000 kg/m^3. Specific weight (w) is weight per unit volume and its basic unit is N/m^3.

Since a mass of 1 kg weighs 9.81 N at the earth's surface, $w = 9.81\rho$, or in more general terms, $w = \rho g$. The specific weight of fresh water is approximately 1000 kg/m^3 $\times$ 9.81 m/s^2 = 9.81 kN/m^3.
The relative density or specific gravity of a substance is the ratio:

$$\frac{\text{density of substance}}{\text{density of pure water}}$$

The pressure at a depth h below the free surface of liquid at rest is given by:
$p = \rho g h = wh$
If the units of ρ, g and h are respectively kg/m^3, m/s^2 and m, the units of p are

$$\text{kg/m}^3 \times \text{m/s}^2 \times \text{m} = \frac{\text{kg} \times \text{m/s}^2}{\text{m}^2} = \frac{\text{N}}{\text{m}^2}$$

This unit is called the pascal (Pa). It is a very small unit and the kilopascal (kPa) is more often used. The pressure given by the formula is the pressure measured above that at the free surface of the liquid. It is usually, therefore, the pressure

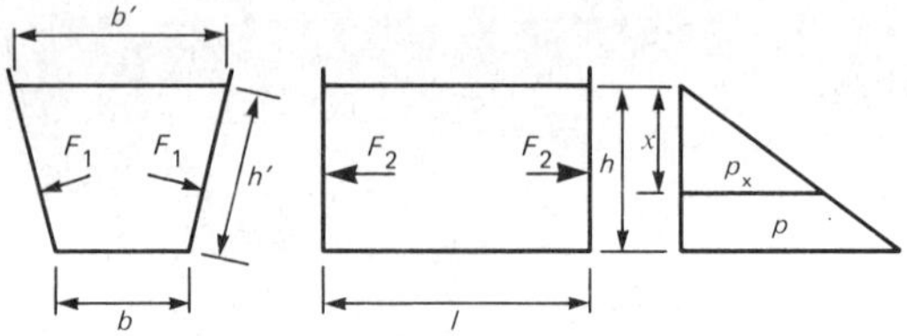

(a)

(b)
Variation of pressure

Fig 1

measured above atmospheric pressure. This pressure is called gauge pressure. Another unit of pressure, used in particular for atmospheric pressures, is the bar. 1 bar = 10^5 N/m^2 = 100 kPa.
The pressure at the base of the tank shown in *Fig 1(a)* is:
$p = \rho g h = wh.$
The pressure increases to this value uniformly from zero at the free surface. The pressure variation is shown in *Fig 1(b).* At any intermediate depth x the pressure is
$p_x = \rho g x = wx.$

It may be shown that the average pressure on any wetted plane surface is the pressure at the centroid, the centre of area. The sloping sides of the tank of *Fig 1* are rectangular and therefore the average pressure on them is the pressure at half depth:

$$\frac{p}{2} = \frac{\rho g h}{2} = \frac{wh}{2}$$

The force on a sloping side is the product of this average pressure and the area of the sloping side:

$$F_1 = \frac{\rho g h}{2} \times lh' = \frac{whlh'}{2}$$

where h' is the slant height. The pressure and consequently the forces F_1 are at right angles to the sloping sides as shown in *Fig 1*. The average pressure on the vertical trapezoidal ends of the tank is not the pressure at half depth. This is because the centroid of an end is not at half depth – it is rather higher. The depth of the centroid is given by:

$$\frac{h}{3}\left(\frac{2b + b'}{b + b'}\right) \text{ (see } \textit{Fig 1}\text{)}$$

The average pressure on an end is therefore:

$$\frac{\rho g h}{3}\left(\frac{2b + b'}{b + b'}\right) = \frac{wh}{3}\left(\frac{2b + b'}{b + b'}\right)$$

The forces F_2 on the vertical trapezoidal ends of the tank are horizontal forces given by the product of this average pressure and the area of the trapezium,

$$\left(\frac{b + b'}{2}\right)h.$$

The force on the base of the tank is
$\rho g h \times$ area of base $= \rho g h l b = whlb$
For a tank with vertical sides this is the weight of liquid in the tank. (See *Problem 8*)

3 In any vessel containing homogeneous liquid at rest and in continuous contact, the pressure must be the same at all points at the same level. In a U-tube, as shown in *Fig 2,* with the liquid in the lower part at rest, the pressure must be the same on both sides for all levels up to $X_1 X_2$. The pressure at X_1, however, is greater than the pressure at Y_1 by an amount $p_1 = w_1 h = \rho_1 gh$, where w_1, ρ_1 are the specific weight and density respectively of the liquid, or gas, between X_1 and Y_1.

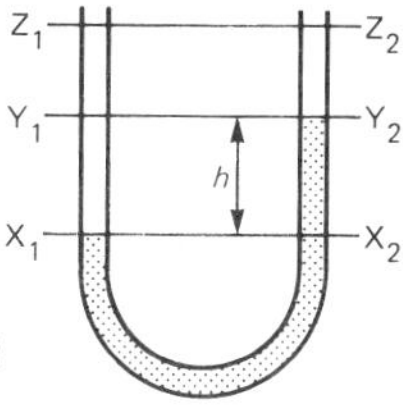

Fig 2

Similarly, the pressure at X_2 is greater than the pressure at Y_2 by an amount given by $p_2 = w_2 h = \rho_2 gh$, where w_2, ρ_2 are the specific weight and density respectively of the liquid in the bottom of the U-tube.

For practical reasons ρ_2 must be greater than ρ_1 and the pressure at Y_1 will exceed that at Y_2 by $p_2 - p_1 = (w_2 - w_1)h = (\rho_2 - \rho_1)gh$. If the upper limbs of the U-tube contain air or any other gas or gas mixture w_1 and ρ_1 can reasonably be ignored, giving

$p_2 - p_1 = w_2 h = \rho_2 gh$

If the upper limbs contain a lighter liquid, then the pressure difference may be expressed as

$$p_2 - p_1 = \rho_1(d - 1)gh = (d - 1)w_1 h \text{ where } d = \frac{\rho_2}{\rho_1}$$

A common arrangement is mercury and water, in which case d is the relative density of mercury approximately 13.6. This gives,

$p_2 - p_1 = 12.6\,\rho gh = 12.6\,wh$, ρ and w being respectively, the density and specific weight of water. The pressure difference at Z_1 Z_2 will be the same as at Y_1 Y_2 if both limbs contain the same liquid between these levels. This follows from the fact that the pressure increase from Z_1 to Y_1 is the same as the increase from Z_2 to Y_2. (See *Problem 9*)

4 **Archimedes' principle**

If a solid body is immersed in a liquid, the apparent loss of weight is equal to the weight of liquid displaced. If V is the volume of the body below the surface of the liquid, then the apparent loss of weight is $W = Vw = V\rho g$, where w, ρ are respectively the specific weight and density of the liquid. If ρ is known and W obtained from a simple experiment, V can be calculated. Hence the density of the solid body can be calculated. If V is known and W obtained, ρ can be calculated. If a body floats on the surface of a liquid all of its weight appears to have been lost. The weight of liquid displaced is equal to the weight of the floating body. (See *Problems 10 and 11*)

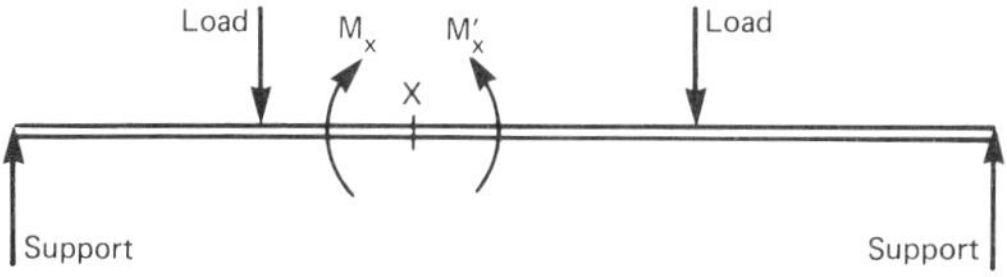

Fig 3

5 **Shearing force and bending moment**

For equilibrium of a beam, the forces to the left of any section such as X in *Fig 3*, must balance the forces to the right. Also the moment about X of the forces to the left must balance the moment about X of the forces to the right.

Although for equilibrium the forces and moments cancel, the magnitude and nature of these forces and moments are important as they determine both the stresses at X and the beam curvature and deflection. The resultant force to the left of X and the resultant force to the right of X, (forces or components of forces transverse to the beam), constitute a pair of forces tending to shear the beam at this section. **Shearing force** is defined as the force transverse to the beam at a given section tending to cause it to shear at that section. By convention, if the tendency is to shear as shown in *Fig 4(a)*, the shearing force is regarded as positive, i.e., $+F$, if the tendency to shear is as shown in *Fig 4(b)*, it is regarded as negative, i.e. $-F$.

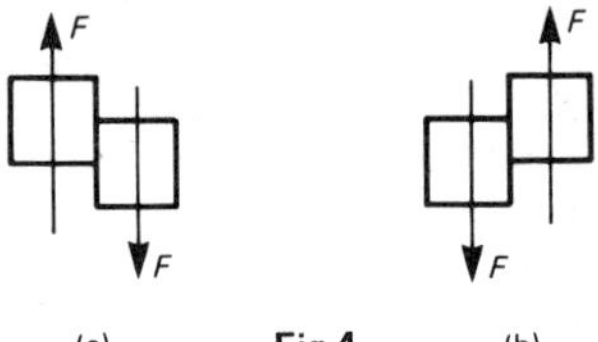

Fig 4

The **bending moment** at a given section of a beam is defined as the resultant moment about that section of either all of the forces to its left – or of all of the forces to its right. In *Fig 3* it is M_X or $M_X{}'$. These moments, clockwise to the left and anticlockwise to the right, will cause the beam to bend concave upwards, called **'sagging'**. By convention this is regarded as positive bending (i.e. the bending moment is a positive bending moment). Where the curvature produced is concave downwards (called **'hogging'**), the bending moment is regarded as negative.

The values of shearing force and bending moment will usually vary along a beam. Diagrams showing the shearing force and bending moement for all sections of a beam are called shearing force and bending moment diagrams respectively.

Shearing forces and shearing force diagrams are less important than bending moments, but can be very useful in giving pointers to the more important bending moment diagrams. For example, wherever the shearing force is zero, the bending moment will be a maximum or a minimum. (See *Problems 12 to 18*)

B. WORKED PROBLEMS ON STATICS AND HYDROSTATICS

STATIC EQUILIBRIUM OF NON-CURRENT FORCES

Problem 1 In *Fig 5*, ABCD is a square. If the four forces are in equilibrium and force $F_1 = 6$ N, calculate the magnitudes of the other forces.

Fig 5

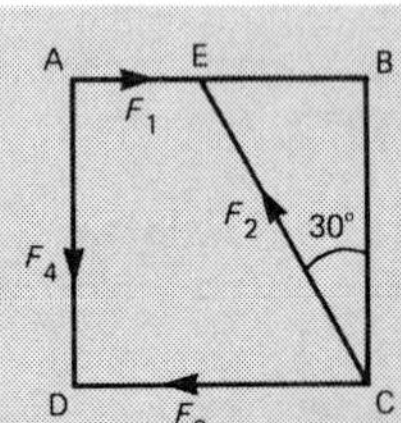

If moments are taken about the point of intersection of any two of the unknown forces, only one unknown force is left to be considered. If the point C is chosen, for example, the moments of F_2 and F_3 about this point are zero and the moment of F_1 must be balanced by the moment of F_4.

i.e., $F_1\ x = F_4\ x$, where x is the side of the square
Hence, $F_4 = F_1 = 6$ N.

The other forces can now be obtained from a force polygon as shown in *Fig 6*, which can be solved by calculation or by drawing. LMNP is a square of which the sides all represent 6 N. From *Fig 6*,

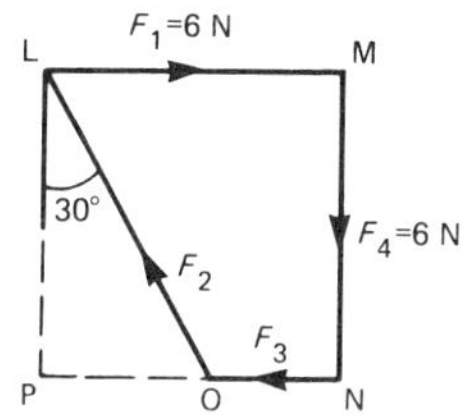

Fig 6

$$\text{force } F_2 = \frac{6\text{ N}}{\cos 30°} = 6.928\text{ N}$$

Also OP represents 6 N × tan 30°, i.e., 3.464 N
F_3, represented by NP – OP, is 6 N – 3.464 N = 2.536 N.
The three unknown forces are therefore:
F_2 = **6.928 N;**
F_3 = **2.536 N;**
F_4 = **6 N**

As an alternative to drawing the force polygon, moments can be taken about the other points of intersection. Taking moments about D, only F_1 and F_2 have to be considered. The lever for F_2 is $x \cos 30° = 0.866\,x$
Hence, $F_1 \times x = F_2 \times 0.866\,x$,

$$\text{from which, } F_2 = \frac{F_1}{0.866} = \frac{6\text{ N}}{0.866} = \mathbf{6.928\ N}$$

Referring to *Fig 5*. BE = BC tan 30° = $0.5774\,x$
AE = $x - 0.5774\,x = 0.4226\,x$

Taking moments about E, only F_3 and F_4 have to be considered.
Hence, F_4 . AE = F_3 . x

$$\text{from which, } F_3 = \frac{F_4 \,.\, \text{AE}}{x} = \frac{6\text{ N} \times 0.4226\,x}{x} = \mathbf{2.536\ N}$$

Problem 2 Solve the previous problem by an entirely graphical method.

The resultant of forces F_1 and F_2, which intersect at point E (space diagram, *Fig 5*) must also pass through E. Similarly, the resultant of forces F_3 and F_4, which intersect at D, must also pass through D. If each of the pairs of forces, F_1 with F_2 and F_3 with F_4 is replaced by its resultant, the system of four forces is reduced to an equivalent system of two forces. These can only be in equilibrium if they are equal and opposite along the same straight line. That is, the resultant of F_1 and F_2 must act along the line ED and the resultant of F_3 and F_4 must also act along this line.

In the vector diagram (*Fig 7*) the vector op, representing the known force, is

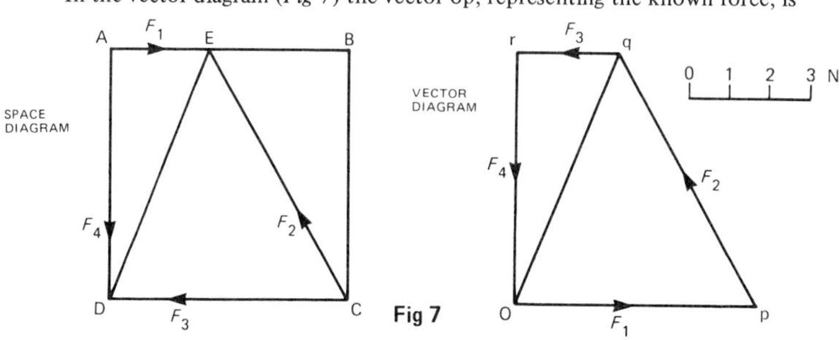

Fig 7

drawn first and then the triangle opq is drawn. This can be done because the direction of pq representing F_2 is known, and also the direction of qo representing the resultant of F_1 and F_2 is known (from the direction of ED in the space diagram).

The directions of F_3 and F_4 being known, the point r is readily located, completing the vector diagram. From this, the magnitudes of F_2, F_3 and F_4 can be measured.

Although moments may appear to have been ignored, they have been allowed for by ensuring that the resultant of any two forces passes through their point of intersection.

It should be noted that for accurate answers by a graphical method, diagrams must be carefully drawn to a good scale. This applies to the space diagram as well as the vector diagram. Even then, answers cannot reasonably be given correct to more than three significant figures.

Problem 3 The bar AD in *Fig 8* is in equilibrium with the four forces shown acting upon it. The distances AB, BC and CD are equal. Calculate the forces F_1, F_2 and F_3.

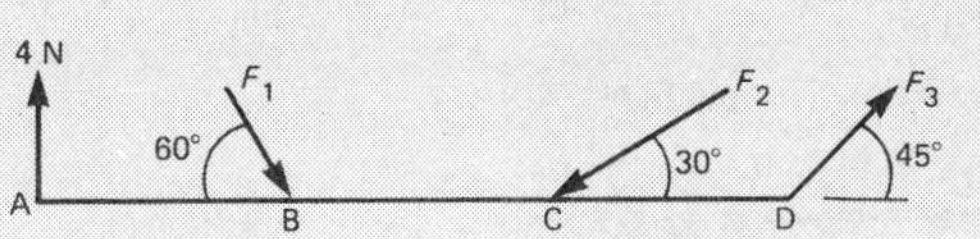

Fig 8

The essential first step is to take moments about one of the points of intersection of two unknown forces. There are several possible approaches after this first step. Forces F_1 and F_2 both have zero moment about point P where they intersect (*Fig 9*).

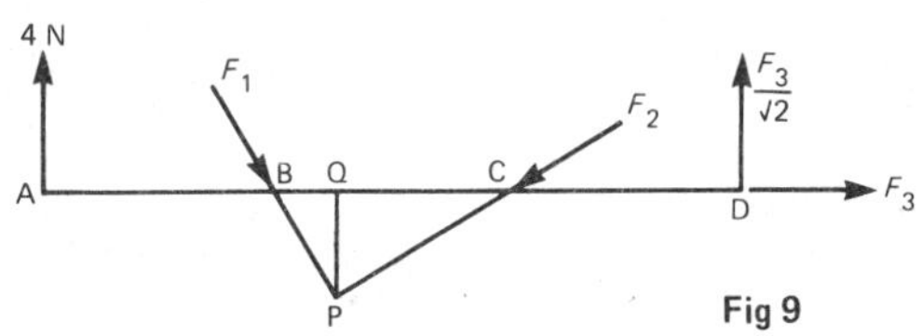

Fig 9

If, therefore, moments are taken about this point, the moment of the known force, 4N, must be balanced by the moment of F_3. In *Fig 9*, BPC, PCQ and BQP are similar triangles, each having angles of 30°, 60° and 90°. Denoting AB = BC = CD by x, then

$BP = BC \sin 30° = x/2$

$BQ = BP \sin 30° = (x/2)\frac{1}{2} = x/4$

Also $PQ = BP \cos 30° = (x/2)\, 0.866 = 0.433\, x$.

The moment of the 4 N force about P is 4 N × distance AQ (the **perpendicular** distance from P to the line of action of the force),

i.e. $4\text{ N} \times (x + \frac{x}{4}) = 5\text{ N}\,x$ (clockwise)

The moment of F_3 is best considered by splitting the force into its components along the bar and at right angles to it. These components are both $F_3/\sqrt{2}$. The moment of the component along AD (away from A) is

$$\frac{F_3}{\sqrt{2}} \times \text{PQ} = \frac{F_3}{\sqrt{2}} \times 0.433x = 0.3062\ F_3x \text{ (clockwise).}$$

The moment of the component at right angles to AD (upwards) is

$$\frac{F_3}{\sqrt{2}} \times \text{QD} = \frac{F_3}{\sqrt{2}} \times 1.75x = 1.2374\ F_3x \text{ (anticlockwise)}$$

Equating clockwise moments about P to anticlockwise moments about P gives:

$5\ \text{N}x + 0.3062\ F_3x = 1.2374\ F_3x$

from which, $F_3 = \dfrac{5\ \text{N}}{1.2374 - 0.3062} = \mathbf{5.369\ N}$

With the number of unknown forces reduced to two, a force polygon can be drawn to scale or sketched and solved by trigonometry. Alternatively, moments can be taken about any convenient point on one of the remaining unknown forces. If moments are taken about B, the lever for F_2 is BP (see *Fig 9*) $= x/2$. The lever for F_3 is $\text{BD}/\sqrt{2}$, i.e. $x\sqrt{2}$. The moments of the forces about B are therefore:

clockwise: $4\ \text{N} \times x = 4\ \text{N}x$

$F_2 \times \text{BP} = F_2x/2$

anticlockwise: $F_3 \times x\sqrt{2} = 5.369\ \text{N} \times x\sqrt{2} = 7.592\ \text{N}x$

Equating clockwise and anticlockwise moments,

$$4\ \text{N}x + \frac{F_2x}{2} = 7.592\ \text{N}x$$

$F_2 = (7.592\ \text{N} - 4\ \text{N})\ 2 = \mathbf{7.184\ N}$

If moments are taken about C the levers for F_1 and F_3 are respectively PC (*Fig 9*) $= 0.866x$ and $\dfrac{x}{\sqrt{2}}$.

Equating clockwise and anticlockwise moments,

$$4\ \text{N} \times 2x = F_1 \times 0.866x + F_3\ \frac{x}{\sqrt{2}}$$

$$= F_1 \times 0.866x + \frac{5.369\ \text{N}x}{\sqrt{2}},$$

$$\text{from which, } F_1 = \frac{8\ \text{N} - \dfrac{5.369\ \text{N}}{\sqrt{2}}}{0.866} = \mathbf{4.853\ N}$$

Problem 4 Solve the previous problem by a purely graphical method.

The space diagram and force diagram are given, to scale, in *Fig 10*. Note that in the space diagram the line PR is the line along which will lie both the resultant of F_1 and F_2 and the resultant of the known force of 4 N and F_3.

To draw the vector diagram the vector ab representing the known force is drawn first. The triangle abc is then completed, the directions of bc and ac both being known. The direction of bc is the direction of the force F_3 which it represents and ac is parallel to PR, the line of action of the resultant, represented by ac, of the known force and F_3. The vector ca represents the resultant of F_1 and F_2, and the triangle cad can now be drawn. The direction of cd is the direction of F_1 and the direction of da is the direction of F_2 and both directions are known.

The known force could have been paired with any of the unknown forces.

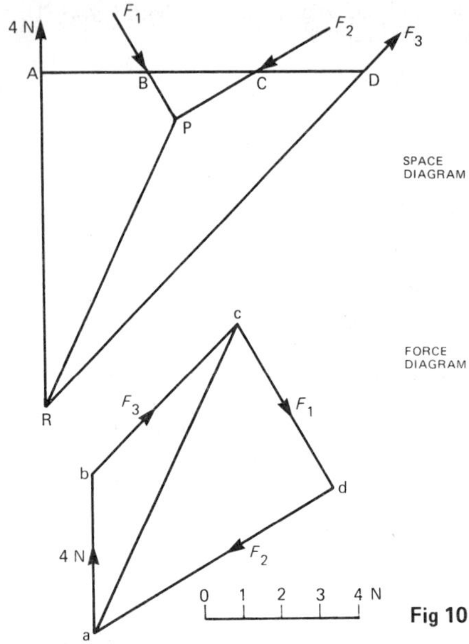

Fig 10

Problem 5 Find the equilibrant of the three forces shown in *Fig 11* by calculation: state its magnitude, direction and position of line of action. The sides of the equilateral triangle ABC are 60 mm long.

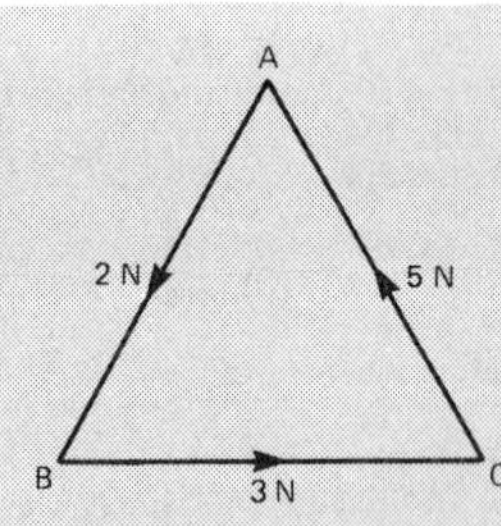

Fig 11

The magnitude and direction of the equilibrant can readily be obtained from a force polygon or by resolving the forces. The position of the line of action requires the principle of moments since the moment of the equilibrant about any point must balance the resultant moment of the given forces.

If components of forces are considered along BC and at right angles to it, only the 2 N and 5 N forces need to be resolved. The components of the 2 N force are:
2 N × cos 60° = 2 N × 0.5 = 1 N (along BC, right to left)
2 N × sin 60° = 2 N × 0.866 = 1.732 N (downwards)
The components of the 5 N force are:
5 N cos 60° = 5 N × 0.5 = 2.5 N (along BC, right to left)
5 N sin 60° = 5 N × 0.866 = 4.33 N (upwards).
The resultant component forces are therefore:
along BC right to left: 1 N + 2.5 N − 3 N = 0.5 N

upwards: 4.33 N − 1.732 N = 2.598 N
The equilibrant is therefore as shown in *Fig 12* and is given by:

$\sqrt{(0.5^2 + 2.598^2)} = \mathbf{2.646\ N}$

The angle θ is given by:

$$\tan\theta = \frac{0.5}{2.598},$$

from which, $\theta = \mathbf{10.89°}$

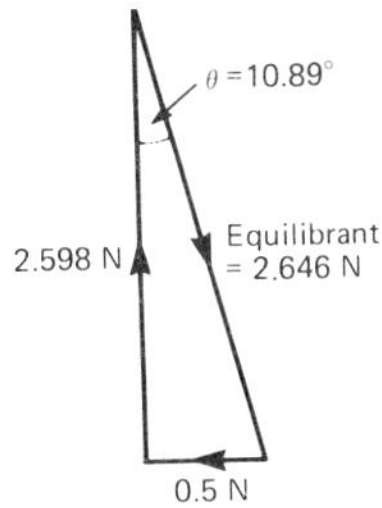

Fig 12

The moment about A of the equilibrant must balance the moment about this point of the 3 N force – the other two given forces have no moment about this point. The moment of the 3 N force is 3 N × (60 mm sin 60°) = 155.9 N mm (anticlockwise).

The moment of the equilibrant about A must be clockwise, the product of its magnitude and the **perpendicular** distance of its line of action from A. To give a clockwise moment the line of action must be to the right of A. The perpendicular distance is $\frac{155.9\ \text{N mm}}{2.646\ \text{N}} = \mathbf{58.92\ mm}$

Problem 6 Solve the previous problem by a purely graphical method.

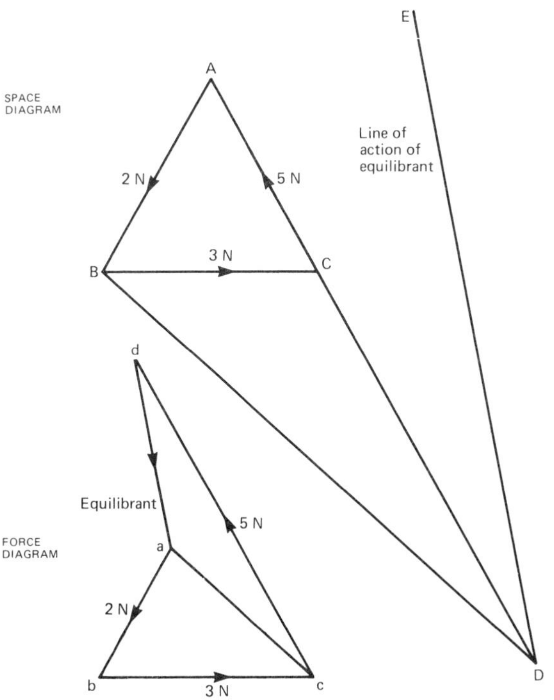

Fig 13

The space diagram and the vector diagram of forces are given in *Fig 13*. The vector diagram can be drawn immediately, with the vector da representing the equilibrium in magnitude and direction.

The vector ac represents the resultant of the 2 N and 3 N forces. As these two forces intersect at B (space diagram) their resultant must also pass through B; otherwise the resultant would have a moment about B whereas the moment of each of the two forces is zero. The line BD in the space diagram, parallel to vector ac, is therefore the line of action of the resultant of the 2N and 3N forces. It intersects the line of action of the 5N force at D.

The original system of three forces has been replaced by an exactly equivalent system of two forces. These two forces intersect at D and their equilibrant must also pass through D. But this equilibrant is also the equilibrant of the original system. The line DE, therefore, parallel to vector da, is the line of action of the equilibrant of the three given forces.

It should again be noted that for accurate results by drawing, great care must be taken over the accuracy of lengths and angles of the diagrams.

Problem 7 The four-wheeled trolley shown in *Fig 14* is to be pulled up the steps of height 20 mm by the horizontal force *F*. The weight *W* of the trolley is 4.14 kN and its centre of gravity is the point G. The diameter of the wheels is 420 mm. What minimum force *F* is required to begin to lift the front wheels off the lower level and up the step? Assume negligible friction in the axles of the trolley.

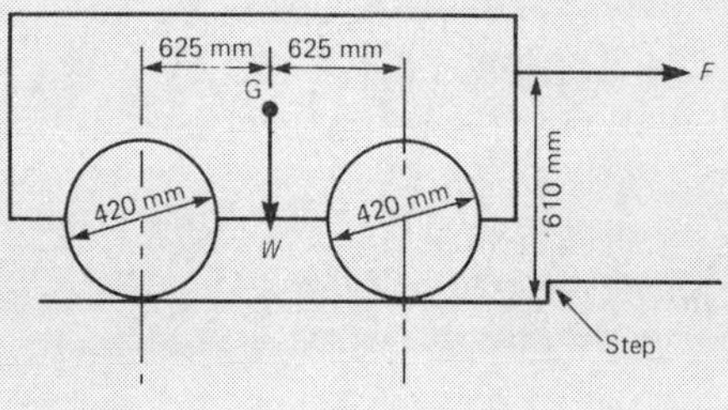

Fig 14

When the front wheels begin to lift they will make contact with the ground only at the edge of the step A, (*Fig 15*).

Because there is no friction in the axles, the unknown supporting forces at A and B will pass through the centres, H and D, of the wheels as shown, intersecting in point C. This is the point about which to take moments.

In triangle ADE,

AD = radius of wheel = 210 mm

ED = (radius of wheel) – (height of step)

= 210 mm – 20 mm = 190 mm.

$$\cos\theta = \frac{ED}{AD} = \frac{190}{210} = 0.90476,$$

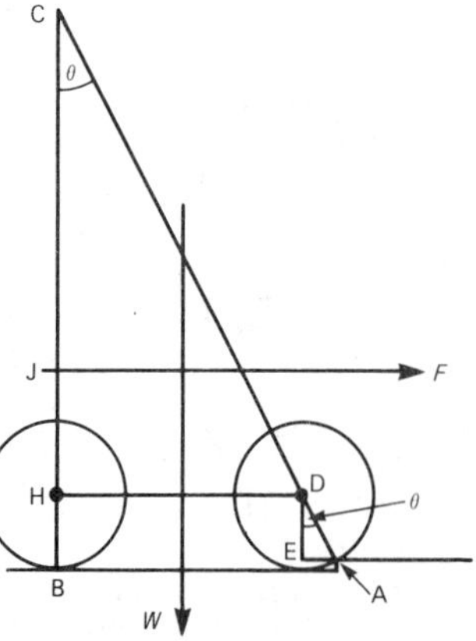

Fig 15

Hence $\theta = 25.209°$, HD = 1.25 m (given) and $\frac{HD}{CH} = \tan\theta$

Hence, $CH = \frac{HD}{\tan\theta} = \frac{1.25 \text{ m}}{\tan 25.209°} = 2.6553$ m.

$$\begin{aligned} CJ &= CH + HB - JB \\ &= 2.6553 \text{ m} + 0.210 \text{ m} - 0.61 \text{ m} \\ &= 2.2553 \text{ m}. \end{aligned}$$

The moment of force F about C is therefore $F \times 2.2553$ m.
The moment of the weight of the trolley about C is
$W \times 0.625 \text{ m} = 4.14 \text{ kN} \times 0.625 \text{ m} = 2.5875$ kN m.
These two moments must be equal for equilibrium as the reactions from the ground pass through C and therefore have no moment about this point,
i.e. $F \times 2.2553 \text{ m} = 2.5875$ kN m

from which, $F = \frac{2.5875 \text{ kN m}}{2.2553 \text{ m}} = \mathbf{1.147\ kN}$

Further problems on static equilibrium of non-concurrent forces may be found in section C(b), Problems 1 to 11, page 46.

HYDROSTATIC PRESSURE

Problem 8 A tank 420 mm long has a horizontal rectangular base, vertical ends and sides sloping at equal angles. It contains water to a depth (h) of 215 mm. The width of the tank at the bottom (b) is 305 mm; the width of the water surface (b') is 375 mm. Calculate the total force on (i) the base; (ii) an end; and (iii) a side of the tank. The depth of the centroid of the wetted area at the end of the tank may be taken as $\frac{h}{3}\left(\frac{2b + b'}{b + b'}\right)$

(i) The pressure at the base of the tank is
$\rho g h = 1000 \text{ kg/m}^3 \times 9.81 \text{ m/s}^2 \times 0.215 \text{ m}$
$= 2.109 \text{ kN/m}^2$ (i.e. kPa).
The force on the base of the tank is therefore
$2.109 \text{ kPa} \times 0.42 \text{ m} \times 0.305 \text{ m} = \mathbf{270.2\ N}$

(ii) The depth of the centroid of a wetted end is:
$$\frac{215 \text{ mm}}{3}\left(\frac{2 \times 305 \text{ mm} + 375 \text{ mm}}{305 \text{ mm} + 375 \text{ mm}}\right) = 103.8 \text{ mm}$$
The average pressure on an end is the pressure at this depth,
i.e. $1000 \text{ kg/m}^3 \times 9.81 \text{ m/s}^2 \times 0.1038 \text{ m} = 1.018 \text{ kN/m}^2$.
The force on an end is therefore given by:
average pressure $\times$ area
$$= 1.018 \text{ kN/m}^2 \times \left(\frac{0.305 \text{ m} + 0.375 \text{ m}}{2}\right) \times 0.215 \text{ m} = \mathbf{74.4\ N}$$

(iii) The average pressure on a sloping side is:
$$\frac{\rho g h}{2} = \frac{2.109 \text{ kN/m}^2}{2} = 1.0545 \text{ kN/m}^2$$

The slant height of a sloping side is:

$$\sqrt{\left[h^2 + \left(\frac{b' - b}{2}\right)^2\right]}$$

$$= \sqrt{\left[(215\ \text{mm})^2 + \left(\frac{375\ \text{mm} - 305\ \text{mm}}{2}\right)^2\right]} = 217.8\ \text{mm}.$$

The force on a sloping side is therefore:
$1.0545\ \text{kN/m}^2 \times 0.42\ \text{m} \times 0.2178\ \text{m} =$ **96.46 N**

Problem 9 The pressure difference between two points on a horizontal pipe carrying water and running full is to be indicated on a U-tube below the pipe. The bottom of the U-tube contains mercury; the rest of it is filled with water. What pressure difference is indicated when the difference in the mercury levels in the U-tube is 310 mm? Take the density of water as 1000 kg/m³ and the relative density of mercury as 13.6.

The pressure difference at level XX′ (*Fig 16*) is the same as the pressure difference between the two points in the pipe.

At level YY′, however, there is no pressure difference, as below this there is a U of mercury at rest. The difference in pressure at XX′ and hence between the points on the pipe is therefore the difference in pressure between that due to a column of mercury 310 mm high (X′ Y′) and that due to a column of water 310 mm high (XY). The pressure difference is therefore

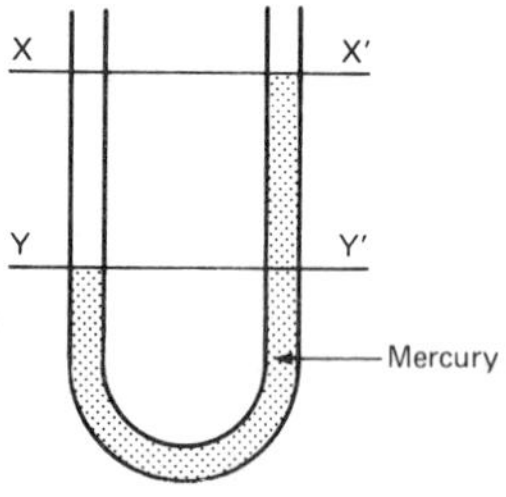

Fig 16

$$310\ \text{mm} \times w_m - 310\ \text{mm} \times w_w$$
$$= 310\ \text{mm}\ (w_m - w_w)$$
$$= 310\ \text{mm}\ (\rho_m g - \rho_w g)$$
$$= 310\ \text{mm}\ \rho_w g(d_m - 1),$$

where w_m = specific weight of mercury;
w_w = specific weight of water;
ρ_m = density of mercury;
ρ_w = density of water;
d_m = relative density of mercury.

Hence, pressure difference $= 0.31\ \text{m} \times 1000\ \text{kg/m}^3 \times 9.81\ \text{m/s}^2\ (13.6 - 1)$
$= 0.31\ \text{m} \times 9.81\ \text{kN/m}^3 \times 12.6 =$ **38.32 kN/m²** (kPa)

Further problems on hydrostatic pressure may be found in section C(b), Problems 12 to 22, page 48.

ARCHIMEDES' PRINCIPLE

Problem 10 A body weighs 2.760 N in air and 1.925 N when completely immersed in water of density 1000 kg/m³. Calculate (a) the volume of the body; (b) the density of the body and (c) the relative density of the body.

(a) The apparent loss of weight is 2.760 N − 1.925 N = 0.835 N.
This is the weight of water displaced, i.e. $V\rho g$,
where V is the volume of the body and ρ is the density of water
i.e. $0.835 \text{ N} = V \times 1000 \text{ kg/m}^3 \times 9.81 \text{ m/s}^2$

$$= V \times 9.81 \text{ kN/m}^3$$

$$\text{Hence, } V = \frac{0.835}{9.81 \times 10^3} \text{ m}^3 = \mathbf{8.512 \times 10^{-5} \text{ m}^3}$$

$$= \mathbf{8.512 \times 10^4 \text{ mm}^3}$$

(b) The density of the body $= \dfrac{\text{mass}}{\text{volume}} = \dfrac{\text{weight}}{g \times V}$

$$= \frac{2.760 \text{ N}}{9.81 \text{ m/s}^2 \times 8.512 \times 10^{-5} \text{ m}^3} = \frac{\frac{2.760}{9.81} \text{ kg} \times 10^5}{8512 \text{ m}^3}$$

$$= \mathbf{3305 \text{ kg/m}^3 = 3.305 \text{ tonne/m}^3}$$

(c) Relative density $= \dfrac{\text{density}}{\text{density of water}}$

The relative density of the body is therefore

$$\frac{3305 \text{ kg/m}^3}{1000 \text{ kg/m}^3} = \mathbf{3.305}$$

Problem 11 A rectangular watertight box is 560 mm long, 420 mm wide and 210 mm deep. It weighs 223 N.
(a) If it floats with its sides and ends vertical in water of density 1030 kg/m³, what depth of the box will be submerged?
(b) If the box is held completely submerged in water of density 1030 kg/m³, by a vertical chain attached to the underside of the box, what is the force in the chain?

(a) The apparent weight of a floating body is zero. That is, the weight of the body is equal to the weight of liquid displaced. This is given by:
$V\rho g$
where V is the volume of liquid displaced, and
ρ is the density of the liquid.
Here, $223 \text{ N} = V \times 1030 \text{ kg/m}^3 \times 9.81 \text{ m/s}^2$
$= V \times 10.104 \text{ kN/m}^3$

$$\text{Hence, } V = \frac{223 \text{ N}}{10.104 \text{ kN/m}^3} = 22.07 \times 10^{-3} \text{ m}^3$$

This volume is also given by $lb\,d$
where l = length of box;
b = breadth of box,
d = depth of box submerged,

i.e. $22.07 \times 10^{-3} \text{ m}^3 = 0.56 \text{ m} \times 0.42 \text{ m} \times d$

$$\text{Hence depth submerged, } d = \frac{22.07 \times 10^{-3}}{0.56 \times 0.42} = 0.09384 \text{ m},$$

$$= \mathbf{93.84 \text{ mm.}}$$

(b) The volume of water displaced is the total volume of the box. The upthrust or buoyancy of the water, i.e., the 'apparent loss of weight', is greater than the weight of the box. The force in the chain accounts for the difference.

Volume of water displaced, $V = 0.56\ \text{m} \times 0.42\ \text{m} \times 0.21\ \text{m} = 4.9392 \times 10^{-2}\ \text{m}^3$
Weight of water displaced $= V\rho g = 4.9392 \times 10^{-2}\ \text{m}^3 \times 1030\ \text{kg/m}^3 \times 9.81\ \text{m/s}^2$
$= 499.1\ \text{N}.$
The force in the chain = weight of water displaced – weight of box
$= 499.1\ \text{N} - 223\ \text{N} =$ **276.1 N.**

Further problems on Archimedes' Principle may be found in section C(b), Problems 23 to 30, page 49.

SHEARING FORCE AND BENDING MOMENT

Problem 12 The beam AB in *Fig 17* is simply-supported at its ends and carries a single concentrated load at C. Calculate the shearing forces and bending moments at intervals of 1 m along the beam.

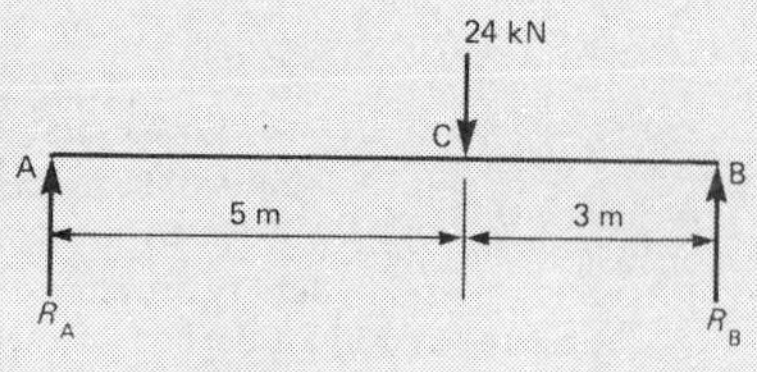

Fig 17

It is first necessary to calculate the reactions at A and B. The beam is simply-supported at A and B which means that it rests on supports at these points giving vertical reactions. The general conditions for equilibrium require that the resultant moment about any point must be zero, and total upward force must equal total downward force. Therefore, taking moments about A, the moment of R_B must balance the moment of the load at C:

$$R_B \times 8\ \text{m} = 24\ \text{kN} \times 5\ \text{m} = 120\ \text{kN m}$$

$$R_B = \frac{120\ \text{kN m}}{8\ \text{m}} = 15\ \text{kN}$$

and $R_A = 24\ \text{kN} - 15\ \text{kN} = 9\ \text{kN}$.

Immediately to the right of A the shearing force is due to R_A and is therefore 9 kN. As this force to the left of the section considered, is upwards, the shearing force is positive. The shearing force is the same for all points between A and C as no other forces come on the beam between these points.

When a point to the right of C is considered, the load at C as well as R_A must be considered, or alternatively, R_B on its own. The shearing force is 15 kN, either obtained from $R_B = 15$ kN, or from load at C – $R_A = 15$ kN. For any point between C and B the force to the right is upwards and the shearing force is therefore negative. It should be noted that the shearing force changes suddenly at C.

The bending moment at A is zero, as there are no forces to its left. At a point 1 m to the right of A the moment of the only force R_A to the left of the point is $R_A \times 1\ \text{m} = 9\ \text{kN} \times 1\ \text{m} = 9\ \text{kN m}$. As this moment to the left is clockwise the bending moment is positive, i.e. it is + 9 kN m. At points 2 m, 3 m, 4 m and 5 m to the right of A the bending moments are respectively:

$R_A \times 2\ \text{m} = 9\ \text{kN} \times 2\ \text{m} = 18\ \text{kN m}$
$R_A \times 3\ \text{m} = 9\ \text{kN} \times 3\ \text{m} = 27\ \text{kN m}$
$R_A \times 4\ \text{m} = 9\ \text{kN} \times 4\ \text{m} = 36\ \text{kN m}$
$R_A \times 5\ \text{m} = 9\ \text{kN} \times 5\ \text{m} = 45\ \text{kN m}$

All are positive bending moments.

For points to the right of C, the load at C as well as R_A must be considered or,

Table 1

Distance from A (m)	0	1	2	3	4	5	6	7	8
Shearing force (kN)	+9	+9	+9	+9	+9	+9 −15	−15	−15	−15
Bending moment (kN m)	0	+9	+18	+27	+36	+45	+30	+15	0

more simply, R_B alone can be used. At points 5 m, 6 m and 7 m from A the bending moments are respectively:

$R_B \times 3$ m = 15 kN $\times$ 3 m = 45 kN m
$R_B \times 2$ m = 15 kN $\times$ 2 m = 30 kN m
$R_B \times 1$ m = 15 kN $\times$ 1 m = 15 kN m

As these moments to the right of the points considered are anti-clockwise they are all positive bending moments. At B the bending moment is zero as there is no force to its right.

The results are summarised in *Table 1*.

Problem 13 Draw to scale the shearing force and bending moment diagrams for the beam shown in *Fig 17*.

Making use of the values calculated in *Problem 12*, the diagrams are as shown (in the usual manner) in *Fig 18*. A stepped shearing force diagram, with horizontal and vertical lines only, is always obtained when the beam carries concentrated loads only. A sudden change in shearing force occurs where the concentrated loads, including the reactions at supports, occur. For this type of simple loading the bending moment diagram always consists of straight lines, usually sloping. Sudden changes of bending moment cannot occur except in the unusual circumstance of a moment being applied to a beam as distinct from a load.

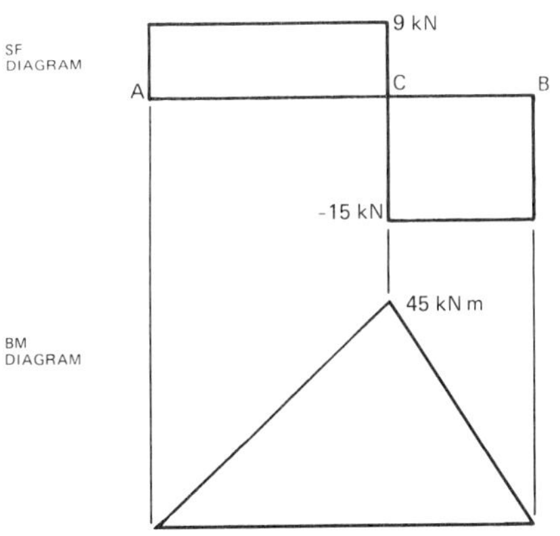

Fig 18

Problem 14 Find for the cantilever shown in *Fig 19*, the shearing forces and bending moments at A, B and C.

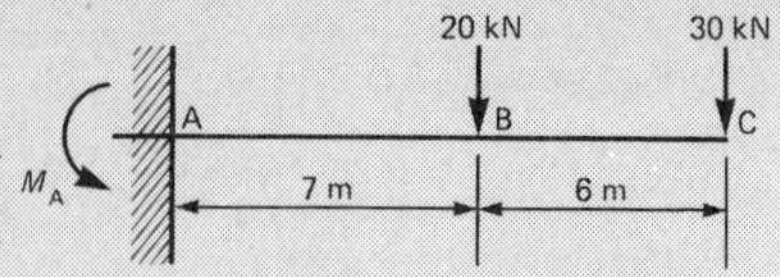

Fig 19

The cantilever must be built into the supporting wall at A, since to satisfy the conditions of equilibrium, the reaction here must not only provide an upward force to balance the downward effect of the two loads, but also a moment to balance the moments about A of the loads.

The only force to the right of any point between B and C is the load at C. This then is the shearing force from a point immediately to the right of B to C. As the force to the right of any point is downwards, the shearing force is positive and is +30 kN. For any point between A and B there is a resultant downward force of 50 kN. The shearing force over this length of the beam is therefore +50 kN. (Note that the shearing force suddenly increases at B.) As there are no forces or moments to the right of C, the bending moment here is zero.

The bending moment at B is the product of the single load to its right, at C, and its distance away:

M_B = 30 kN × 6 m = 180 kN m.

As this is a clockwise moment to the right of the section, the bending moment is negative, a 'hogging' moment. The bending moment at A is the combined moment about this point of the loads at B and C.

M_A = 30 kN × 13 m + 20 kN × 7 m
= 390 kN m + 140 kN m = 530 kN m.

Again this is a negative, 'hogging moment'.

The support at A must provide a reaction moment (anticlockwise as shown in *Fig 19*) of 530 kN m as well as an upward reaction of 50 kN m. The results are summarised in *Table 2*.

Table 2

Position	A	B	C
Shearing force (kN)	+50	+50 +30	+30
Bending moment (kN m)	–530	–180	0

Problem 15 Draw the shearing force and bending moment diagrams for the beam shown in *Fig 19*.

Making use of the values calculated in *Problem 14*, the diagrams are shown in *Fig 20*. Again, the loads being concentrated forces, the shearing force diagram has a stepped appearance, the bending moment diagram consists of sloping straight lines with no sudden changes in value of bending moment except at A, due to the reaction moment at the wall.

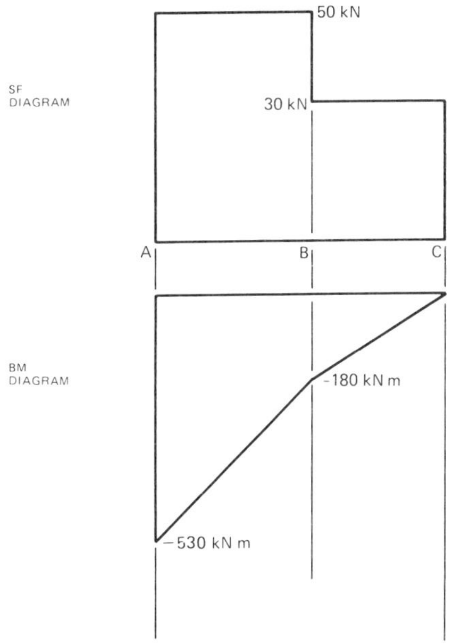

Fig 20

Problem 16 The beam shown in *Fig 21* is simply-supported at its ends and carries a load of 12 kN uniformly distributed over its full span. Find expressions for the shearing force and bending moment at X, a distance of x metres from A. Hence draw the shearing force and bending moment diagrams for the beam.

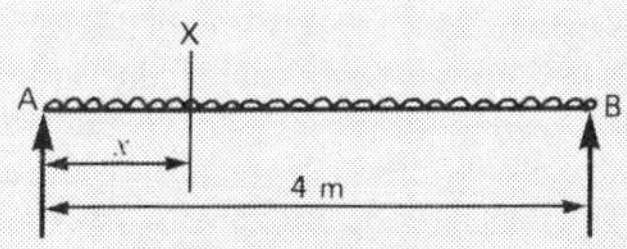

Fig 21

As the beam is symmetrical, the reactions will be equal, 6 kN each. The shearing force at X is more easily considered as the net force to its left. The amount of distributed load on AX is 12 kN $\times \frac{x}{4} = 3x$ kN.

The shearing force at X is therefore $R_A - 3x$ kN = **(6 − 3x) kN.**

The bending moment at X is the difference between the moments about this point, of R_A and the length x m of distributed load. This length of distributed load has its centre of gravity $\frac{x \text{ m}}{2}$ from X. Its moment about X is therefore $3x$ kN $\times \frac{x \text{ m}}{2}$ $= \frac{3x^2}{2}$ kN m.

The bending moment at X is therefore: $R_A \times x - \frac{3x^2}{2}$

$$= \mathbf{(6x - \frac{3x^2}{2})\ kN\ m.}$$

The expression for shearing force is the equation of a straight line. Giving x values of 0, 1, 2, 3 and 4 gives values for shearing force of 6 kN, 3 kN, 0, –3 kN and –6kN respectively. The expression for bending moment is the equation of a parabola. Giving x values of 0, 1, 2, 3 and 4 gives values for bending moment of; 0, 4.5 kN m, 6 kN m, 4.5 kN m and 0, respectively.

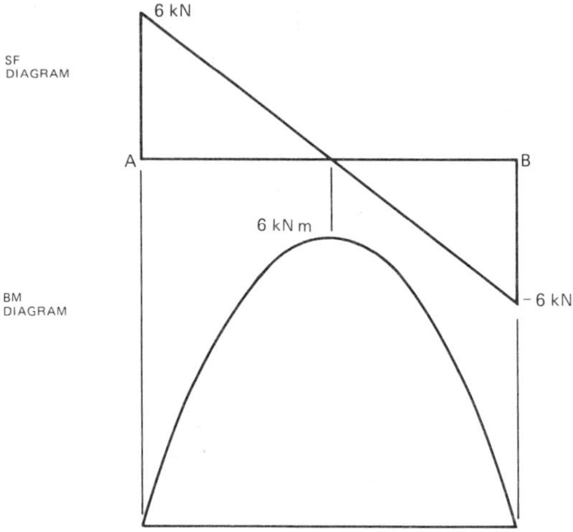

Fig 22

These values are all positive and are symmetrical about the centre of the beam.

The shearing force and bending moment diagrams are given in *Fig 22*. Note the sloping straight line of the shearing force diagram and the curve of the bending moment diagram. These are features always found with uniformly distributed loads. Note also that the maximum bending moment occurs where the shearing force is zero.

Problem 17 The beam AB, 16 m long, is simply-supported at C and D as shown in *Fig 23*. In addition to concentrated loads at A and E it carries a uniformly distributed load of 2 kN/m over a 10 m length extending from E to B. Calculate the shearing force and bending moment at 1 m intervals along the beam.

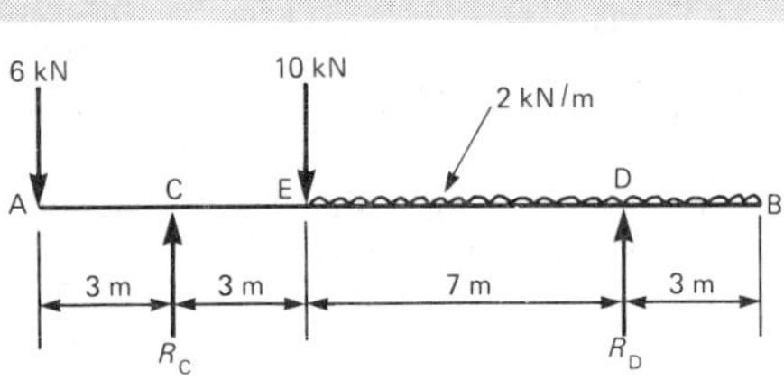

Fig 23

The total distributed load is 20 kN with its centre of gravity half-way between E and B, i.e., 5 m from E and 8 m from C. Therefore, taking moments about C to find the reaction at D:

$$R_D \times 10 \text{ m} = 10 \text{ kN} \times 3 \text{ m} + 20 \text{ kN} \times 8 \text{ m} - 6 \text{ kN} \times 3 \text{ m}$$
$$= 30 \text{ kN m} + 160 \text{ kN m} - 18 \text{ kN m} = 172 \text{ kN m}$$

Hence, $R_D = \dfrac{172 \text{ kN m}}{10 \text{ m}} = 17.2$ kN.

R_C = total load $- R_D$
$= 6 \text{ kN} + 10 \text{ kN} + 20 \text{ kN} - 17.2 \text{ kN}$
$= 36 \text{ kN} - 17.2 \text{ kN} = 18.8 \text{ kN}.$

Shearing forces

The shearing force for all points between A and C is 6 kN as this is the only force to the left. As it is downwards the SF is negative.

The shearing force for all points between C and E is $R_C - 6$ kN, i.e. 12.8 kN, positive. The shearing force at a point immediately to the right of E is:
$R_C - 6 \text{ kN} - 10$ kN, i.e., $18.8 \text{ kN} - 16 \text{ kN} = 2.8$ kN, positive.
To find the shearing force at a point x m to the right of E (up to D) the weight of x m of distributed load, i.e, $2x$ kN must be allowed for as well as R_C and the loads at A and E. That is, the shearing force is
$R_C - 6 \text{ kN} - 10 \text{ kN} - 2x \text{ kN} = (2.8 - 2x)$ kN.

Giving x values of 1, 2, 3, 4, 5, 6 and 7 gives values for the shearing force of 0.8 kN, −1.2 kN, − 3.2 kN, − 5.2 kN, −7.2 kN, −9.2 kN and −11.2 kN respectively. The last value applies to a point immediately to the left of D. For a point immediately to the right of D, the shearing force changes by an amount equal to R_D. It is therefore 17.2 kN − 11.2 kN = 6 kN.

This value is confirmed if forces are considered to the right of a point x metres from B (up to D). As the distributed load is 2 kN/m the shearing force at a point distant x m from B is $2x$ kN. For values of x: 1, 2, and 3 (point D), the shearing force is respectively 2 kN, 4 kN and 6 kN. These are positive shearing forces as the downward force is to the right.

Bending moments

Consider a point x metres to the right of A. For values of x up to 3 the bending moment is entirely due to the load at A and is negative:
$M_x = -6x$ kN m

Giving x values of 1, 2 and 3 m gives values of bending moment
$M_1 = -6$ kN m
$M_2 = -12$ kN m
$M_3 = -18$ kN m

For values of x in the range 3 to 6, R_C must be considered as well as the load at A:
$M_x = R_C\,(x - 3) - 6x$
$= (18.8x - 56.4 - 6x)$ kN m
$= (12.8x - 56.4)$ kN m

Giving values 3, 4, 5 and 6 to x gives values of bending moment:
$M_3 = (38.4 - 56.4)$ kN m $= -18$ kN m (as before)
$M_4 = (51.2 - 56.4)$ kN m $= -5.2$ kN m
$M_5 = (64 - 56.4)$ kN m $= +7.6$ kN m
$M_6 = (76.8 - 56.4)$ kN m $= +20.4$ kN m
If now a point is considered y metres from B, a length y m of distributed load, amounting to $2y$ kN, must be considered, with its centre half-way along, $y/2$ m from B and $y/2$ m from the point under consideration. The moment of this load is therefore $2y \text{ kN} \times y/2 \text{ m} = y^2$ kN m. For values of y up to 3 this will give the bending moment, negative as it is a clockwise moment to the right, causing hogging.

Giving values 1, 2 and 3 to y gives bending moment values of -1 kN m, -4 kN m and -9 kN m respectively, the last value being the bending moment at D.

For values of y from 3 to 10, the same expression applies for the effect of the distributed load. Here, however, the effect of R_D must be included: R_D $(y-3)$m = $(17.2y-51.6)$ kN m. This is a positive effect. The expression then for for the bending moment for points between E and D is: $(17.2y - 51.6 - y^2)$ kN m, where y is the distance in metres from B. Giving values 4, 5, 6, 7, 8, 9 and 10 to y gives values for bending moment of:

$68.8 - 51.6 - 16 = 1.2$ kN m
$86 - 51.6 - 25 = 9.4$ kN m
$103.2 - 51.6 - 36 = 15.6$ kN m
$120.4 - 51.6 - 49 = 19.8$ kN m
$137.6 - 51.6 - 64 = 22.0$ kN m
$154.8 - 51.6 - 81 = 22.2$ kN m
$172.0 - 51.6 - 100 = 20.4$ kN m

This last value is the bending moment at E and confirms an earlier value.
The values are summarised in *Table 3*.

Table 3

Distance from A (m)	0	1	2	3	4	5	6	7	8	9	10
Shearing force (kN)	−6	−6	−6	−6 12.8	12.8	12.8	2.8 12.8	0.8	−1.2	−3.2	−5.2
Bending moment (kN m)	0	−6	−12	−18	−5.2	7.6	20.4	22.2	22.0	19.8	15.6

Distance from A (m)	11	12	13	14	15	16
Shearing force (kN)	−7.2	−9.2	−11.2 6	4	2	0
Bending moment (kN m)	9.4	1.2	−9	−4	−1	0

Problem 18 Draw to scale the shearing force and bending moment diagrams for the beam shown in *Fig 23*. Calculate the position and magnitude of the maximum sagging moment in the beam.

Making use of the values calculated in *Problem 17*, the diagrams are shown in *Fig 24*.

Features of the shearing force diagram to be noted are:

(i) sloping lines where distributed loads occur;
(ii) sudden changes of shearing force where concentrated loads occur – the change being equal to the concentrated load.

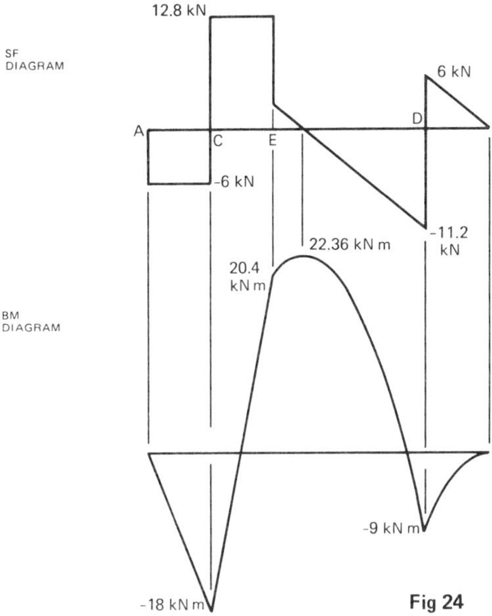

Fig 24

Features of the bending moment diagram to be noted are:

(i) curves where distributed loads occur,

(ii) sloping straight lines elsewhere,

(iii) negative, i.e. hogging moments due to loads on the over-hanging ends of the beam,

(iv) no sudden changes in values of bending moment, but sudden changes in slope where concentrated forces (including the reactions) occur.

The maximum bending moment values occur where the shearing force is zero. The shearing force changes from positive to negative at C and D. At these points the bending moment has maximum negative, hogging, values.

The shearing force also passes through a zero value to the right of E where the bending moments are sagging moments. The shearing force is 0.8 kN at a point 1 m to the right of E and is dropping by 2 kN/m. It is therefore zero at a further distance of 0.8/2 m, i.e. 0.4 m from E.

The sagging moment is a maximum at a distance of 7.4 m from A.

Its value is given by $(17.2y - 51.6 - y^2)$ kN m (see solution to *Problem 17*) where y is the distance from B in metres (i.e. 8.6 m). That is, the maximum sagging moment is:

$(17.2 \times 8.6 - 51.6 - 8.6^2)$ kN m = (147.92 − 51.6 − 73.96) kN m

= **22.36 kN m.**

Further problems on shearing force and bending moment may be found in section C(b) following, Problems 31 to 46, page 50.

C. FURTHER PROBLEMS ON STATICS AND HYDROSTATICS

(a) SHORT ANSWER PROBLEMS

1 State what is meant by static equilibrium.
2 What three conditions must be met for a body to be in static equilibrium?
3 Define density and state its basic unit.
4 State what is meant by relative density.
5 Pressure is the ratio of $\frac{\ldots\ldots\ldots\ldots}{\ldots\ldots\ldots\ldots}$ and its unit is the ....
6 The pressure at depth h below the free surface of a liquid is given by ....
7 State the principle of Archimedes.
8 State what is meant by a shearing force.
9 Explain what is meant by a concentrated load.
10 Explain what is meant by a distributed load.
11 State what is meant by a bending moment.
12 When comparing shear force and bending moment diagrams, what is the significance of the value of the bending moment when the shear force is zero?

(b) CONVENTIONAL PROBLEMS

Static equilibrium of non-concurrent forces

1 The four co-planar forces shown in *Fig 25* are in equilibrium. If the force F_1 is 6 N, find the magnitudes of the other forces.
[F_2 = 2.828 N; F_3 = 5.657 N, F_4 = 6 N]

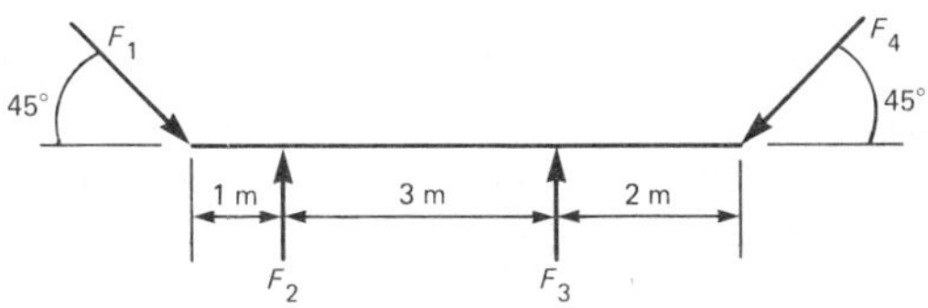

Fig 25

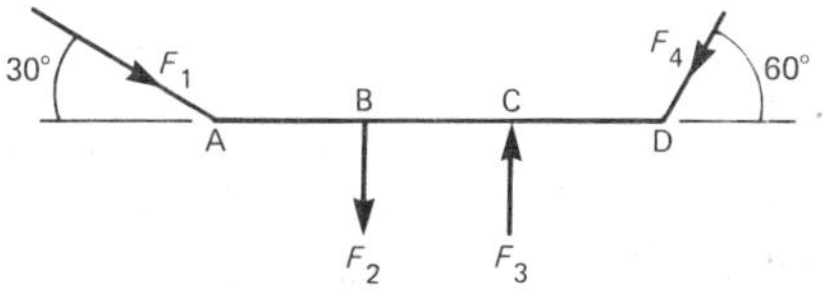

Fig 26

2 The four co-planar forces shown in *Fig 26* are in equilibrium. The lengths AB, BC and CD are equal. If force F_2 is 8 N, calculate the magnitudes of the other forces.
[F_1 = 16 N, F_3 = 40 N, F_4 = 27.71 N]

3 The four co-planar forces shown in *Fig 27* are in equilibrium. The lengths AB, BC and CD are equal. If the force F_1 is 4 N, calculate the magnitudes of the other forces.
[F_2 = 4.308 N; F_3 = 8.352 N; F_4 = 5.72 N]

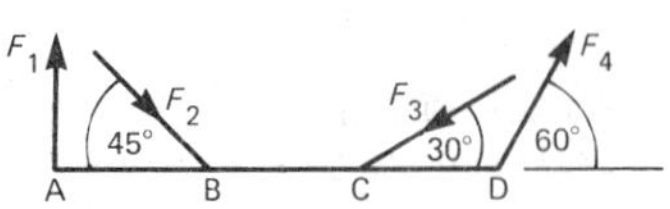

Fig 27

4 Solve *Problem 3* entirely by drawing.

5 The four co-planar forces shown in *Fig 28* are in equilibrium. The equilateral triangle ABC has sides of 1 m; the distance CD is 250 mm. If force F_1 is 7 N, find the magnitudes of the other forces.
[F_2 = 7 N; F_3 = 24.25 N; F_4 = 21 N]

Fig 28

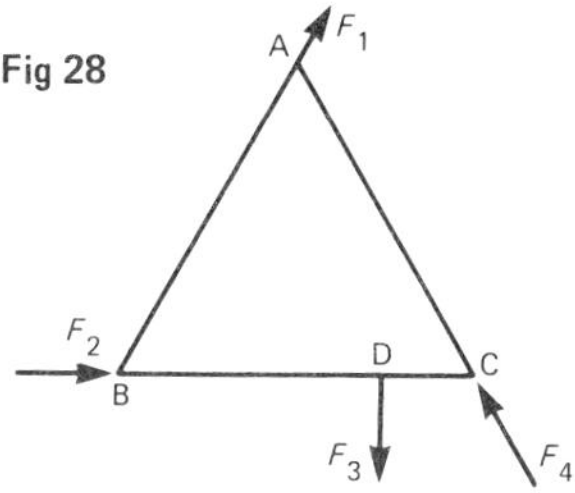

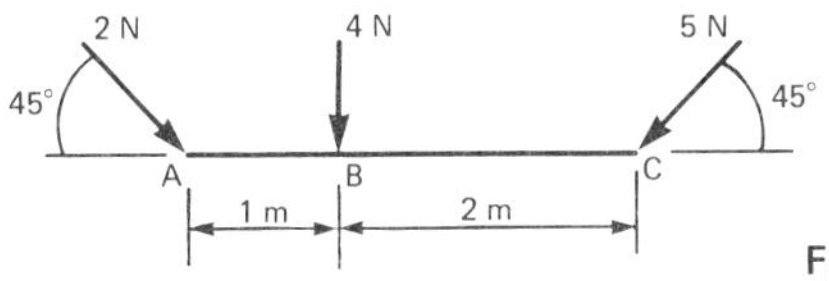

Fig 29

6 Find the resultant of the forces shown in *Fig 29*: state its magnitude, the angle which it makes with the direction AC, and the position of the intersection of its line of action with the line AC.
[9.198 N; 103.33° clockwise; 1.632 m to the right of A]

7 In *Fig 30*, ABCD is a square of side 400 mm. Calculate the resultant of the four forces shown: state its magnitude and direction and the position of the intersection of its line of action with CD (produced if necessary).
[3.162 N; 71° 57′ anticlockwise from CD, 400 mm to the right of C]

8 Solve *Problem 7* graphically.

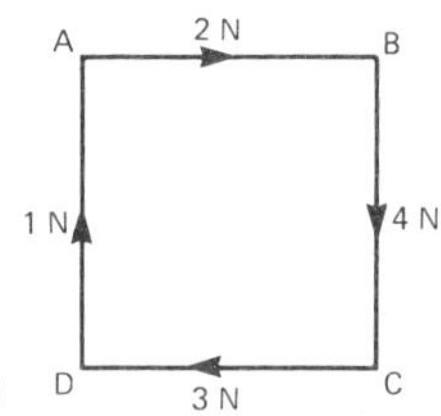

Fig 30

9 The boom AB in *Fig 31* is hinged at A and supported by the rope BC. The boom has a weight of 2 kN and its centre of gravity is at its mid-point. Find the force in the rope, and the force at the hinge (magnitude and direction) when a load of 12 kN is carried at B.
[17.76 kN, 16.21 kN at 18.42° above horizontal]

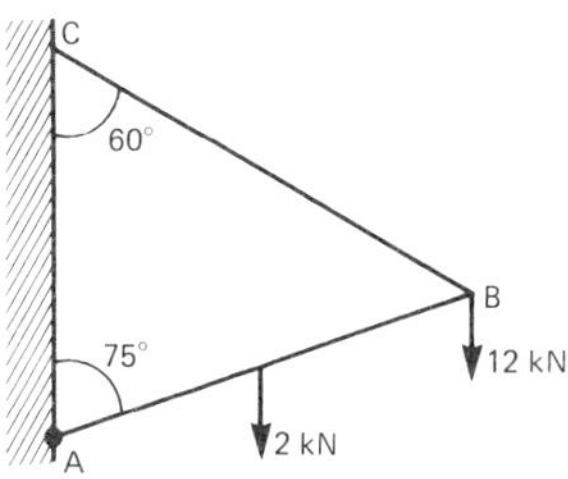

Fig 31

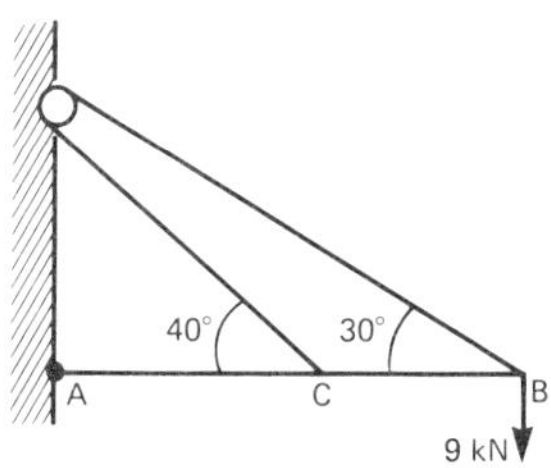

Fig 32

10 The boom AB (*Fig 32*), hinged at A, is supported horizontally by a rope attached to it at B and at C, the mid-point of AB. The angles between the rope and the boom are 30° at B and 40° at C. The rope passes over a frictionless pulley so that the tension is the same at B and at C. Find the tension in the rope and the reaction at A (magnitude and direction) when a load of 9 kN is carried at B.
[10.96 kN; 18.23 kN, 11.14° below the horizontal]

11 The weight W of the four-wheeled trolley shown in *Fig 33* is 2.4 kN. Its centre of gravity is the point G. The rear axle is seized up so that the wheels slide along, the coefficient of friction being 0.3. The front wheels turn freely. Find the minimum force F required to move the trolley.

[327 N]

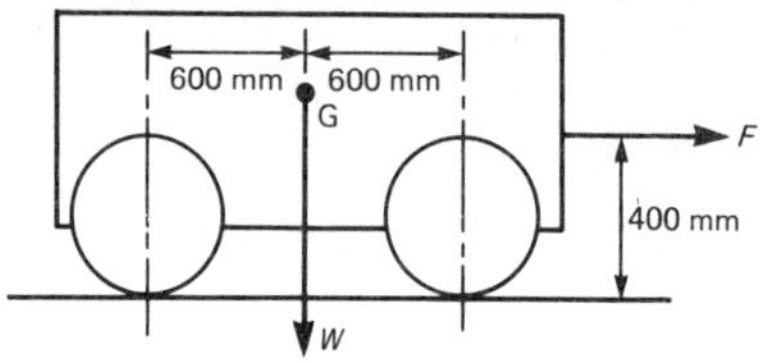

Fig 33

Hydrostatic pressure

(Take the density of water as 1000 kg/m^3)

12 Calculate the pressure at a depth of 20 m below the surface of water at rest.

[196.2 kPa]

13 Calculate the pressure at a depth of 2.4 m below the surface of oil in a tank open to atmosphere at the top. Take the relative density of the oil as 0.8.

[18.84 kPa]

14 A rectangular tank contains oil of relative density 0.82 to a depth of 1.4 m. In the enclosed space above the oil the gauge pressure is 12 kPa. What is the absolute pressure at the bottom of the tank? Take atmospheric pressure as 101 kPa.

[124.3 kPa]

15 A rectangular tank with vertical sides is 600 mm long and 400 mm wide. The tank contains water to a depth of 520 mm. What are the total forces on the base of the tank, a side and an end due to hydrostatic pressure? Take the specific weight of water as 9.81 kN/m^3.

[1.224 kN; 795.8 N; 530.5 N]

16 A tank has a horizontal rectangular base, 460 mm long and 285 mm wide, vertical sides, and ends sloping at 45° outwards. It contains water to a depth of 182 mm. Calculate the total force on (a) the base; (b) a side; and (c) an end of the tank. The depth of the centroid of the wetted area of a side of the tank may be taken as

$$\frac{h}{3}\left(\frac{2l + l'}{l + l'}\right),$$

where h is the depth of water;

l is the length of the base of the tank;

l' is the length of the free surface of the water.

[234.1 N; 94.45 N; 65.48 N]

17 The pressure inside an air vessel is indicated by a U-tube containing water. One side of the U-tube is connected to the vessel and the other is open to atmosphere. What is the gauge pressure of the air in the vessel when the water level on the side of the U-tube open to atmosphere is 124 mm higher than the level on the other side?

[1.216 kN/m^2 or 1.216 kPa]

18 What is the gauge pressure in the air vessel of *Problem 17* when the water level on the side of the U-tube open to atmosphere is 116 mm lower than the level on the other side?

[−1.138 kN/m^2 or −1.138 kPa]

19 If the atmospheric pressure is 102 kPa, what is the absolute pressure in the air vessel for the conditions given in *Problem 18*?

[100.86 kPa]

20 Water flows along a horizontal pipe, filling the pipe. A U-tube is connected across two tapping points A and B of the pipe as in *Fig 34*. The bottom of the U-tube contains mercury; the rest of it is filled with water. What is the pressure difference between A and B when the difference in the mercury levels is 85.6 mm? Take the densities of water and mercury as 1 tonne/m^3 and 13.6 tonne/m^3 respectively.

[10.58 kN/m^2 or 10.58 kPa]

21 Water flows along a horizontal pipe, filling the pipe. A U-tube is connected across two tapping points A and B as shown in *Fig 34*. The bottom of the U-tube is to contain a heavier liquid; the rest of it will be filled with water. The maximum pressure difference between A and B to be recorded is 4.2 kPa and this is to give a difference in levels of the heavier liquid of 400 mm. What is the relative density required for the heavier liquid?

[2.070]

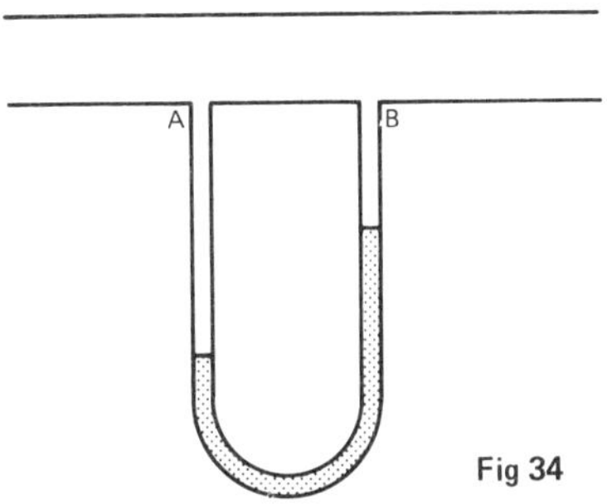

Fig 34

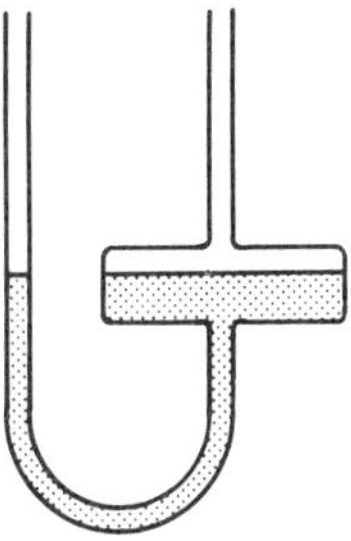

Fig 35

22 The U-tube shown in *Fig 35* contains mercury in the lower part. Above this it is filled with water on both sides; the two limbs are connected to tapping points in a horizontal pipe conveying water. The diameter of the tubing used is 20 mm but on one side, as shown, there is an expanded section of 200 mm diameter. The mercury level on this side is within the expanded section. What pressure difference is recorded by a movement of the mercury level in the expanded section of 4.5 mm from its mean position? Take the relative density of mercury as 13.6.

[56.18 kPa]

Archimedes' Principle

23 A body of volume 0.124 m^3 is completely immersed in water of density 1000 kg/m^3. What is the apparent loss of weight of the body?

[1.216 kN]

24 A body of weight 27.4 N and volume 1240 mm^3 is completely immersed in water of specific weight 9.81 kN/m^3. What is its apparent weight?

[15.24 N]

25 A body weighs 512.6 N in air and 256.8 N when completely immersed in oil of density 810 kg/m^3. What is the volume of the body?

[32.19 dm^3 = 0.03219 m^3]

26 A body weighs 243 N in air and 125 N when completely immersed in water. What will it weigh when completely immersed in oil of relative density 0.8?

[148.6 N]

27 A watertight rectangular box, 1.2 m long and 0.75 m wide, floats with its sides and ends vertical in water of density 1000 kg/m³. If the depth of the box in the water is 280 mm, what is its weight?

[2.575 kN]

28 A body weighs 18 N in air and 13.7 N when completely immersed in water of density 1000 kg/m³. What are the density and relative density of the body?

[4.186 tonne/m³; 4.186]

29 A watertight rectangular box is 660 mm long, and 320 mm wide. Its weight is 336 N. If it floats with its sides and ends vertical in water of density 1020 kg/m³, what will be its depth in the water?

[159 mm]

30 A watertight drum has a volume of 0.165 m³ and a weight of 115 N. It is completely submerged in water of density 1030 kg/m³, held in position by a single vertical chain attached to the underside of the drum. What is the force in the chain?

[1.552 kN]

Shearing force and bending moment

31 A beam AB, 6 m long and simply supported at its ends, carries three concentrated loads: 4 kN at C, 1 m from A; 6 kN at D, 2.5 m from A, and 3.6 kN at E, 4.5 m from A. Calculate the shearing forces:

(a) between A and C;
(b) between C and D;
(c) between D and E;
(d) between E and B.

[7.733 kN; 3.733 kN; 2.867 kN, 5.867 kN]

32 A beam AB, 7 m long and simply supported at its ends carries three concentrated loads: 4.5 kN at C, 1.5 m from A; 6 kN at D, 4 m from A, and 8 kN at E, 5.5 m from A. Calculate the shearing forces

(a) between A and C;
(b) between C and D;
(c) between D and E;
(d) between E and B.

[7.82 kN; 3.32 kN; 2.68 kN; 10.68 kN]

33 A beam AB, 8 m long, is simply supported at its ends. It carries a uniformly distributed load of 1.5 kN/m over a 4 m length of the beam, extending from a point 1 m from A to a point 5 m from A. Find the shearing forces at points distant 2 m, 4 m and 6 m from A.

[2.25 kN; 0.75 kN; 2.25 kN]

34 A beam AB, 6 m long, is simply supported at its ends. It carries a concentrated load of 4 kN at a point 2.5 m from A and also a uniformly distributed load of 1.2 kN/m over a 4 m length of the beam, extending from a point 1.5 m from A to a point 5.5 m from A. Find the shearing forces at points distant 1 m, 2 m, 3 m, 4 m and 5 m from A.

[4.333 kN; 3.733 kN, 1.467 kN; 2.667 kN; 3.867 kN]

35 A beam AB, simply supported at its ends, is 9 m long. It carries a uniformly distributed load of 1 kN/m over its entire length and two concentrated loads: 5 kN at a point 4 m from A and 6 kN at a point 6 m from A. Find the distance from A of a point where the shearing force is zero. Find also the magnitude and position of the maximum shearing force on the beam.

[4.278 m; 10.72 kN at B]

36 A beam AB, 8 m long, is simply supported at its ends. It carries a uniformly distributed load of 1.6 kN/m over a length AC of the beam, 3 m long and a uniformly distributed load of 1.2 kN/m over the remaining 5 m, from C to B. Find the distance from A of a point where the shearing force is zero.

[3.813 m]

37 A beam AB, 7 m long and simply supported at its ends carries two concentrated loads: 4 kN at C, 2 m from A, and 5 kN at D, 4 m from A. Calculate the bending moments at C and D.

[10 kN m; 12 kN m]

38 A beam AB, 7 m long, is simply supported at A and at C, 5 m from A. It carries three concentrated loads: 3 kN at D, 2 m from A; 4 kN at E, 4 m from A, and 2 kN at B. Calculate the bending moments at C, D and E.

[4 kN m; 3.6 kN m; 1.2 kN m]

39 A beam AB, 6 m long, is simply supported at its ends. It carries a uniformly distributed load of 2 kN/m over a length of 3 m, from a point 1 m from A to a point 4 m from A. It also carries two concentrated loads: 3 kN at a point 1.2 m from A and 4 kN at a point 5.1 m from A. Calculate the bending moments at points 1 m, 2 m, 3 m, 4 m and 5 m from A.

[6.5 kN m; 1.5 kN m; 0.5 kN m; 2.5 kN m; 2.5 kN m]

40 A beam AB, simply supported at its ends, is 8 m long. It carries a uniformly distributed load of 2.2 kN/m over its entire length and also a concentrated load of 5 kN at a point 2 m from A. Find the position and magnitude of the maximum bending moment in the beam.

[3.614 m from A; 22.92 kN m]

41 A beam AB, 5.6 m long, is simply-supported at C and D, 600 mm and 5.2 m from A. It carries a uniformly distributed load of 1.8 kN/m over its entire length and also two concentrated loads – 1.5 kN at a point 1 m from A and 1.6 kN at a point 4.6 m from A. Find the position and magnitude of the maximum bending moment in the beam causing sagging.

[2.965 m from A, 5.311 kN m]

42 A beam AB, 7.5 m long, is simply-supported at its ends. It carries three concentrated loads: 2 kN at C, 1.2 m from A; 4.2 kN at D, 4 m from A and 3.2 kN at E, 6 m from A. Draw to scale the shearing force and bending moment diagrams for the beam.

[See *Fig 36*]

43 A cantilever AB, 10 m long, is built in at A. It carries concentrated loads of 15 kN at B and 10 kN at C, 5 m from A; it also carries a uniformly distributed load of 3 kN/m over the 5 m from C to B. Draw to scale the shearing force and bending moment diagrams.

[See *Fig 37*]

44 A beam AB, 8 m long, is simply-supported at its ends. It carries a concentrated load of 24 kN at C, 3 m from A, and a uniformly distributed load of 4 kN/m over the 5 m from C to B. Draw to scale the shearing force and bending moment diagrams.

[See *Fig 38*]

45 A beam AB is 14 m long and is simply-supported at A and at C, 10 m from A. It carries concentrated loads of 16 kN at B and 18 kN at D, 6 m from A; it also carries a uniformly distributed load of 2 kN/m over the 10 m from A to C. Draw to scale the shearing force and bending moment diagrams. Calculate the position and magnitude of the maximum sagging moment.

[See *Fig 39*; 29.16 kN m, 5.4 m from A]

46 A beam AB, 15 m long, is simply-supported at C and D, 3 m and 12 m respectively from A. It carries concentrated loads of 10 kN at B and 15 kN at E, 8 m from A. It also carries a uniformly distributed load of 2 kN/m over the 8 m from A to E. Draw to scale the shearing force and bending moment diagrams.

[See *Fig 40*]

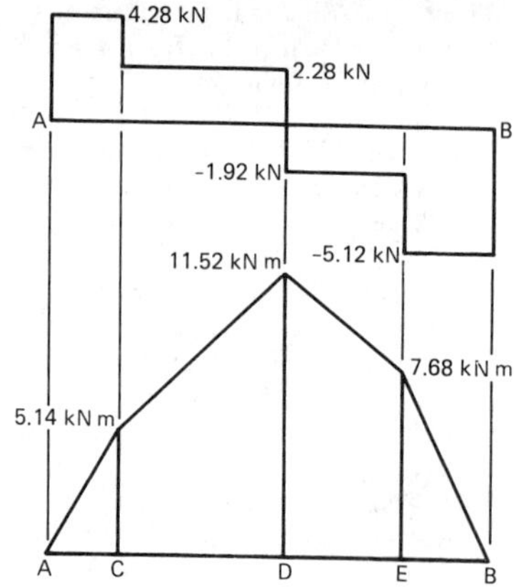

Fig 36

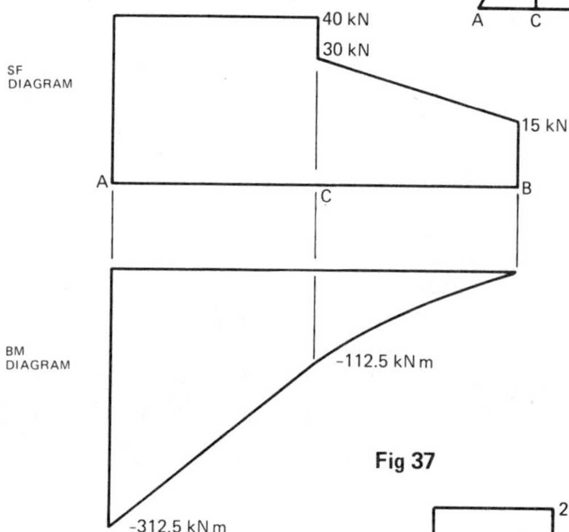

Fig 37

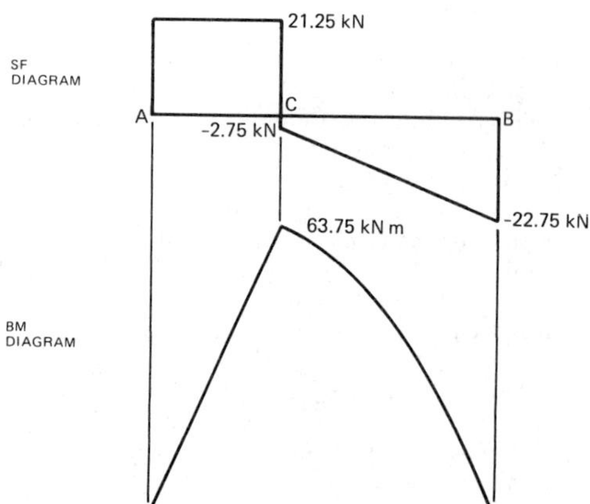

Fig 38

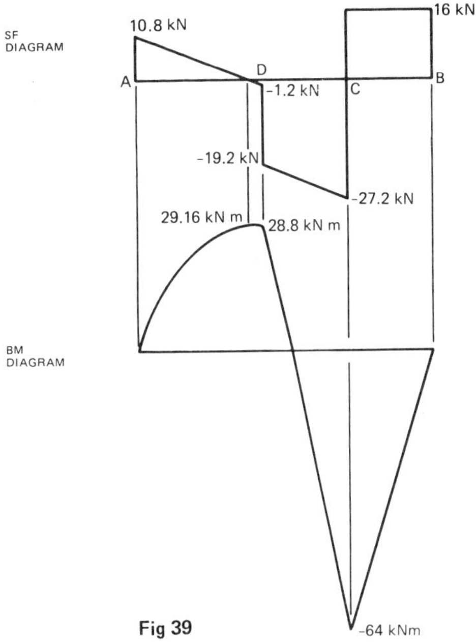

SF DIAGRAM
10.8 kN
16 kN
A
D
C
B
-1.2 kN
-19.2 kN
-27.2 kN
29.16 kN m
28.8 kN m
BM DIAGRAM
-64 kNm

Fig 39

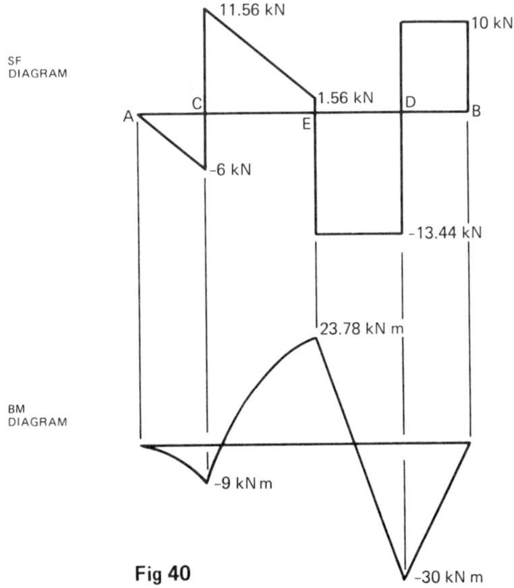

11.56 kN
10 kN
SF DIAGRAM
A
C
1.56 kN
D
B
E
-6 kN
-13.44 kN
23.78 kN m
BM DIAGRAM
-9 kN m
-30 kN m

Fig 40

4 Electric motor applications

A. MAIN POINTS CONCERNED WITH ELECTRIC MOTOR APPLICATIONS

1 A machine is a system having an input and an output

2 (i) When the input to an electrical machine is electrical energy, (seen as applying a voltage to the electrical terminals of the machine), and the output is mechanical energy, (seen as a rotating shaft), the machine is called an **electric motor.** Thus an electric motor converts electrical energy into mechanical energy.
(ii) When the input to an electrical machine is mechanical energy, (seen as, say, a diesel motor coupled to the machine by a shaft), and the output is electrical energy, (seen as a voltage appearing at the electrical terminals of the machine), the machine is called a **generator**. Thus, a generator converts mechanical energy to electrical energy. It can be seen that a motor and a generator are the same type of system, but with inputs and outputs reversed.

3 The efficiency of an electrical machine is the ratio of the output power to the input power and is usually expressed as a percentage. The Greek letter, 'eta', 'η' is used to signify efficiency and since the units are $\frac{\text{power}}{\text{power}}$, then efficiency has no units. Thus

$$\textbf{efficiency, } \eta = \frac{\textbf{output power}}{\textbf{input power}} \times 100\%$$

4 **The action of a commutator on a d.c. motor**
In an electric motor, conductors rotate in a uniform magnetic field. A single-loop conductor mounted between permanent magnets is shown in *Fig 1*. A voltage is applied at points A and B in *Fig 1(a)*. A force, F, acts on the loop due to the interaction of the magnetic field of the permanent magnets and the magnetic field created by the current flowing in the loop. This force is proportional to the flux density, B, the current flowing, I, and the effective length of the conductor, l, i.e., $F = BIl$. The force is made up of two parts, one acting vertically downwards due to the current flowing from C to D and the other acting vertically upwards due to the current flowing from E to F (from Fleming's left hand rule). If the loop is free to rotate, then when it has rotated through 180°, the conductors are as shown in *Fig 1(b)*. For rotation to continue in the same direction, it is necessary for the current flow to be as shown in *Fig 1(b)*, i.e. from D to C and from F to E. This apparent reversal in the direction of current flow is achieved by a process called **commutation**.

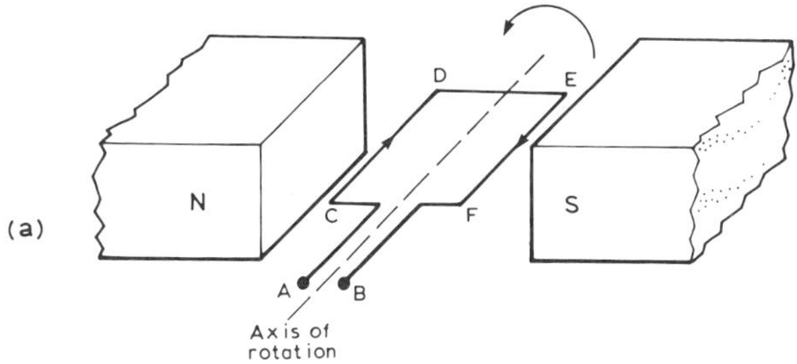

N
S
D
E
C
F
A
B
(a)
Axis of
rotation

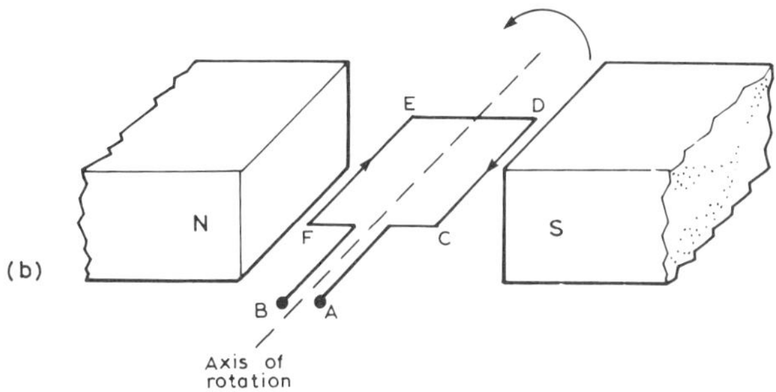

N
S
E
D
F
C
B
A
(b)
Axis of
rotation

Fig 1

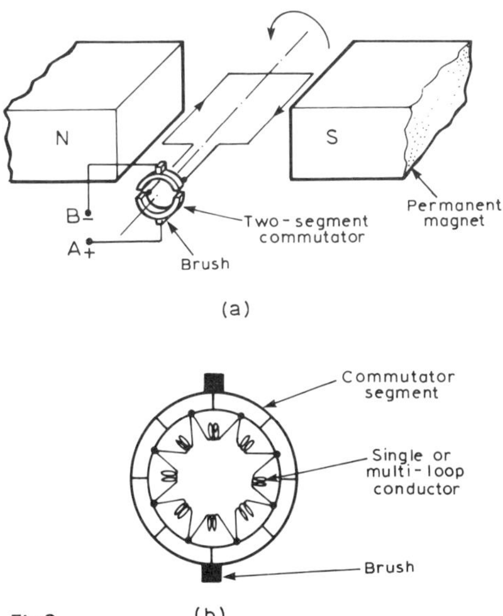

N
S
B−
A+
Two-segment
commutator
Brush
Permanent
magnet
(a)
Commutator
segment
Single or
multi-loop
conductor
Brush
(b)

Fig 2

With reference to *Fig 2(a),* when a direct voltage is applied at A and B, then as the single-loop conductor rotates, current flow will always be away from the commutator for the part of the conductor adjacent to the N-pole and towards the commutator for the part of the conductor adjacent to the S-pole. Thus the forces act to give continuous rotation in an anti-clockwise direction.

The arrangement shown in *Fig 2* is called a 'two-segment' commutator and the voltage is applied to the rotating segments by stationary **brushes**, (usually carbon blocks), which slide on the commutator material, (usually copper), when rotation takes place.

In practice, there are many conductors on the rotating part of a d.c. machine and these are attached to many commutator segments. A schematic diagram of a multi-segment commutator is shown in *Fig 2(b).*

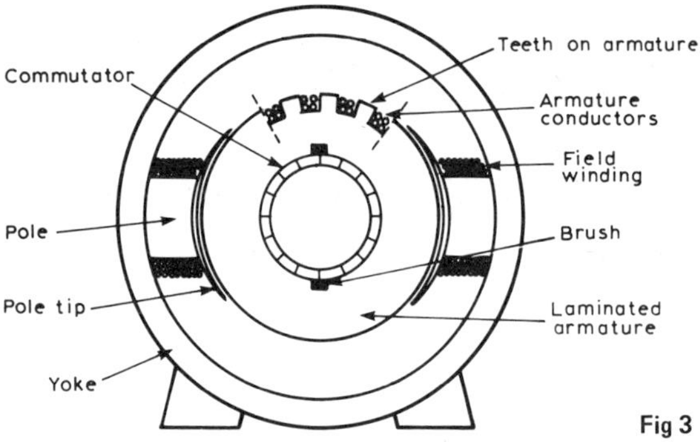

Fig 3

5 **D.C. motor construction**

The basic parts of any d.c. motor are shown in *Fig 3*, and comprise:

(a) a stationary part called the **stator** having,

(i) A steel ring called the **yoke**, to which are attached

(ii) the magnetic **poles**, around which are the

(iii) **field windings**, i.e. many turns of a conductor wound round the pole core. Current passing through this conductor creates an electromagnet, (rather than the permanent magnets shown in *Figs 1 and 2*).

(b) a rotating part called the **armature** mounted in bearings housed in the stator and having,

(iv) a laminated cylinder of iron or steel called the **core**, on which teeth are cut to house the

(v) **armature winding**, i.e. a single or multi-loop conductor system and

(vi) the **commutator**, (see para 4).

6 The average e.m.f. induced in a single conductor on the armature of a d.c. machine is given by

$$\frac{\text{flux cut/rev}}{\text{time of 1 rev}} = \frac{2p\,\Phi}{1/n}$$

where p is the number of **pairs** of poles, Φ is the flux in Wb entering or leaving a pole and n is the speed of rotation in rev/s. Thus the average e.m.f. per conductor is $2\,p\,\Phi n$ volts. If there are Z conductors connected in series, the average e.m.f. generated is $2p\,\Phi nZ$ volts. For a given machine the number of pairs of poles p and the number of conductors connected in series Z are constant, hence the generated

e.m.f. is proportional to Φn. But $2\pi n$ is the angular velocity, ω, in rad/s, hence the generated e.m.f. E is proportional to Φ and to ω, i.e.
generated e.m.f., $E \propto \Phi\omega$ (1)

7 The power on the shaft of a d.c. motor is the product of the torque and the angular velocity, i.e.
shaft power = $T\omega$ watts
where T is the torque in N m and ω is the angular velocity in rad/s. The power developed by the armature current is $E I_a$ watts, where E is the generated e.m.f. in volts and I_a is the armature current in amperes. If losses are neglected then $T\omega = E I_a$. But from para 6, $E \propto \Phi\omega$,
Hence, $T\omega = \Phi\omega I_a$, i.e. $T \propto \Phi I_a$ (2)

8 When the field winding of a d.c. motor is connected in parallel with the armature, as shown in *Fig 4(a)*, the motor is said to be **shunt** wound. If the field winding is connected in series with the armature, as shown in *Fig 4(b)*, then the motor is said to be **series** wound.

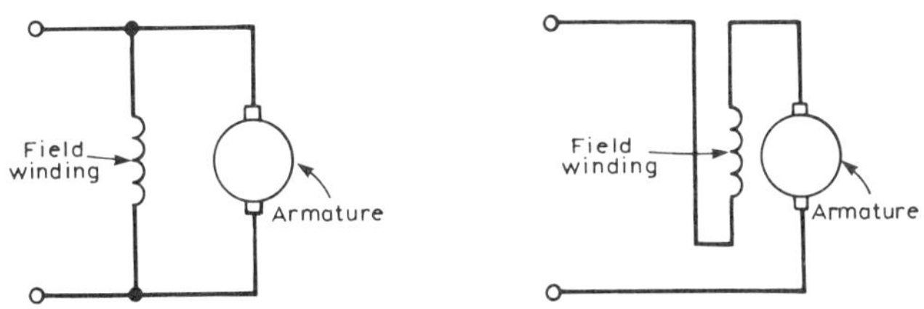

(a) Shunt-wound machine (b) Series-wound machine

Fig 4

9 Depending on whether the motor is series wound or shunt wound, it behaves differently when a load is applied. The behaviour of a d.c. motor under various conditions is shown by means of graphs, called characteristic curves or just **characteristics**. The characteristics shown in the paras below and in the worked problems are theoretical, since they neglect the effects of such things as armature reaction and demagnetising ampere-turns, which are beyond the scope of this text.

10 **Shunt-wound motor characteristics**
The two principal characteristics are the torque/armature current and speed/armature current relationships. From these, the torque/speed relationship can be derived.

(i) The theoretical torque/armature current characteristic can be derived from the expression $T \propto \Phi I_a$, (see para. 7). For a shunt-wound motor, the field winding is connected in parallel with the armature circuit and thus the applied voltage gives a constant field current, i.e., a shunt-wound motor is a constant flux machine. Since Φ is constant, it follows that $T \propto I_a$, and the characteristic is as shown in *Fig 5(a)*.

(ii) The armature circuit of a d.c. motor has resistance due to the armature winding and brushes, R_a ohms, and when armature current I_a is flowing through it, there is a voltage drop of $I_a R_a$ volts. In *Fig 5(b)* the armature resistance is shown as a separate resistor in the armature circuit to help understanding. Also, even though the machine is a motor, because conductors are rotating in a magnetic field, a voltage, $E \propto \Phi\omega$, is generated by the

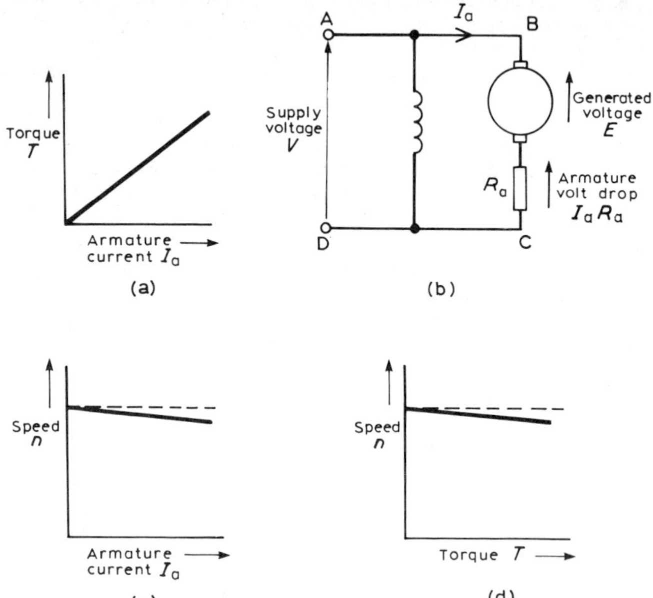

Fig 5

armature conductors. By applying Kirchhoff's voltage law to the armature circuit ABCD in *Fig 5(b),* the voltage equation is
$V = E + I_a R_a$, i.e. $E = V - I_a R_a$.

But from para 6, $E \propto \Phi n$, hence $n \propto \frac{E}{\Phi}$, i.e.

speed of rotation, $n \propto \frac{E}{\Phi} \propto \frac{V - I_a R_a}{\Phi}$ (3)

For a shunt motor, V, Φ and R_a are constants, hence as armature current I_a increases, $I_a R_a$ increases and $V - I_a R_a$ decreases, and the speed is proportional to a quantity which is decreasing and is as shown in *Fig 5(c).* As the load on the shaft of the motor increases, I_a increases and the speed drops slightly. In practice, the speed falls by about 10% between no-load and full-load on many d.c. shunt-wound motors. Due to this relatively small drop in speed, the d.c. shunt-wound motor is taken as basically being a constant-speed machine and may be used for driving lathes, lines of shafts, drilling machines and so on.

(iii) Since torque is proportional to armature current, (see (i) above), the theoretical speed/torque characteristic is as shown in *Fig 5(d).*

11 **Series-wound motor characteristics**
The torque/current, speed/current and speed/torque characteristics are discussed in *Problem 7,* page 63 and the characteristics are shown in *Fig 6.*

12 **Compound-wound motors**
A compound-wound motor has both a series and a shunt field winding, (i.e., one winding in series and one in parallel with the armature), and is usually wound to

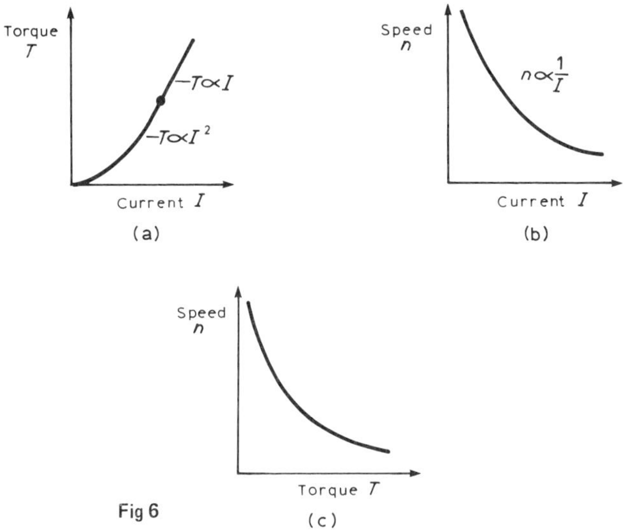

Fig 6

have a characteristic similar in shape to a series wound motor, see *Fig 6*. A limited amount of shunt winding is present to restrict the no-load speed to a safe value. However, by varying the number of turns on the series and shunt windings and the directions of the magnetic fields produced by these windings (assisting or opposing), families of characteristics may be obtained to suit almost all applications. Generally, compound-wound motors are used for heavy duties, particularly in applications where sudden heavy load may occur such as for driving plunger pumps, presses, geared lifts, conveyors, hoists and so on.

13 Two principal types of a.c. motors are in widespread use, these being **induction motors** and **synchronous motors**. Each of these types may be further subdivided into those being run from a three-phase industrial supply, called 3-phase motors and those largely in domestic use, called single-phase motors.

14 When a 3-phase supply is connected to the stator winding on the yoke of a **3-phase induction motor**, a magnetic field of constant magnitude and rotating at constant speed is generated within the stator. When a closed conductor loop mounted on bearings is placed in this magnetic field, the conductors are cut by the rotating field causing a current to flow in the closed loop. The magnetic fields due to the conductor current and the stator interact to create a force on the conductor loop, causing rotation of the conductor loop. In practice, many closed loop conductors are mounted on a cylindrical rotor. The characteristics of a 3-phase induction motor can be altered at the design stage, but frequently the characteristics are very similar to those of a d.c. shunt motor shown in *Fig 5*.

Three-phase induction motors are widely used in industry and constitute almost all industrial drives where a nearly constant speed is required, from small workshops to the largest industrial enterprises.

15 The majority of 'fractional horse-power' motors in domestic use for driving refrigerators, hot water pumps, fans, hair dryers and so on are single-phase induction motors. A single-phase supply connected to the stator winding only

produces a pulsating magnetic field rather than the rotating magnetic field of a 3-phase supply and facilities such as additional windings have to be built-in to the stators of these motors to give some torque at zero speed. There are several devices used and motors are named after the particular device used. Some of these include
'split-phase start',
'capacitor start',
'capacitor start and run',
'shaded pole',
'permanent-split capacitor induction motors', and so on.

16 **3-phase synchronous motors** normally have two windings, a 3-phase stator winding to produce a rotating magnetic field and a rotor winding supplied by a direct current to magnetise the rotor. The rotor is locked magnetically to the rotating magnetic field and for a constant frequency supply, the rotor runs at a constant speed which is directly proportional to the supply frequency.

These motors may be used on many of the loads driven by induction motors but

(i) are normally more expensive;
(ii) require both a.c. and d.c. supplies;
(iii) require additional equipment to run them up to near to their normal running speed, so that the rotor magnetic field can be locked to the rotating magnetic field, and
(iv) have higher maintenance costs than induction motors.

However, 3-phase synchronous motors can run at higher efficiencies and when used adjacent to several induction motors can lead to a higher efficiency of the whole system, (power factor correction).

Due to the disadvantages listed above, three-phase synchronous motors are usually used for applications requiring a large power input, where savings due to higher efficiency outweigh the disadvantages. Their uses include driving large water pumps (power stations, water supply), driving rolling lines in steel mills and driving mine ventilating fans.

17 As for single-phase induction motors, **single-phase synchronous motors** require additional starting devices. They also require a direct current supply to the rotor or a permanent magnet built into the rotor. The former is rarely used and the principal uses of single-phase synchronous motors are largely limited to electric clocks, timing devices, record players and so on, these motors being called 'shaded-pole motors'. As for 3-phase synchronous motors, they run at constant speed when connected to a constant frequency supply.

B. WORKED PROBLEMS ON ELECTRIC MOTOR APPLICATIONS

Problem 1 A 200 V d.c. motor develops a shaft torque of 15 N m at 1200 rev/min. If the efficiency is 80%, determine the current supplied to the motor.

From para 3, the efficiency of a motor = $\frac{\text{output power}}{\text{input power}} \times 100\%$.

The output power of a motor is the power available to do work at its shaft and is given by $T\omega$ or $(T)(2\pi n)$ watts, where T is the torque in N m and n is the speed of rotation in rev/s. The input power is the electrical power in watts supplied to the motor, i.e. VI watts.

Thus for a motor, efficiency $\eta = \left\{\frac{(T)(2\pi n)}{VI}\right\} 100\%$

i.e. $$80 = \left\{\frac{(15)(2\pi)(\frac{1200}{60})}{(200)(I)}\right\} (100)$$

Thus the current supplied, $I = \frac{(15)(2\pi)(20)(100)}{(200)(80)} = \mathbf{11.8\ A}$

Problem 2 A 100-V d.c. generator supplies a current of 15 A when running at 1500 rev/min. If the torque on the shaft driving the generator is 12 N m determine (a) the efficiency of the generator and (b) the power loss in the generator.

(a) From para 3,

the efficiency of a generator $= \frac{\text{output power}}{\text{input power}} \times 100\%$.

The output power is the electrical output, i.e. VI watts. The input power to a generator is the mechanical power in the shaft driving the generator, i.e. $T\omega$ or $(T)(2\pi n)$ watts, where T is the torque in N m and n is speed of rotation in rev/s. Hence, for a generator

efficiency, $\eta = \left\{\frac{VI}{(T)(2\pi n)}\right\} 100\%$

i.e. $$\eta = \left\{\frac{(100)(15)(100)}{(12)(2\pi)(\frac{1500}{60})}\right\}$$

i.e. **efficiency = 79.6%.**

(b) The input power = output power + losses

Hence, $(T)(2\pi n) = VI +$ losses

i.e. losses $= (T)(2\pi n) - VI$

$= (12)(2\pi)(\frac{1500}{60}) - (100)(15)$

i.e. **power loss** = 1885 − 1500 = **385 W**

Problem 3 A d.c. shunt-wound generator running at constant speed generates a voltage of 150 V at a certain value of field current. Determine the change in the generated voltage when the field current is reduced by 20%, assuming the flux is proportional to the field current.

The generated e.m.f. E of a generator is proportional to $\Phi\omega$, (see para. 6), i.e., is proportional to Φn, where Φ is the flux and n is the speed of rotation. It follows that $E = k\Phi n$, where k is a constant. At speed n_1 and flux Φ_1, $E_1 = k\Phi_1 n_1$. At speed n_2 and flux Φ_2, $E_2 = k\Phi_2 n_2$.

Thus, by division: $\frac{E_1}{E_2} = \frac{k\Phi_1 n_1}{k\Phi_2 n_2} = \frac{\Phi_1 n_1}{\Phi_2 n_2}$

The initial conditions are $E_1 = 150$ V, $\Phi = \Phi_1$ and $n = n_1$. When the flux is reduced by 20%, the new value of flux is $\frac{80}{100}$ or 0.8 of the initial value, i.e. $\Phi_2 = 0.8\Phi_1$.

Since the generator is running at constant speed, $n_2 = n_1$. Thus:

$$\frac{E_1}{E_2} = \frac{\Phi_1 n_1}{\Phi_2 n_2} = \frac{\Phi_1 n_1}{0.8\Phi_1 n_1} = \frac{1}{0.8}$$

that is, $E_2 = 150 \times 0.8 = 120$ V.

Thus, a reduction of 20% in the value of the flux **reduces the generated voltage to 120 V** at constant speed.

Problem 4 A 200-V d.c. shunt-wound motor has an armature resistance of 0.4Ω and at a certain load has an armature current of 30 A and runs at 1350 rev/min. If the load on the shaft of the motor is increased so that the armature current increases to 45 A, determine the speed of the motor, assuming the flux remains constant.

The relationship $E \propto \Phi n$ applies to both generators and motors.

For a motor, $E = V - I_a R_a$, (see para 10). Hence,

$$E_1 = 200 - 30 \times 0.4 = 188 \text{ V, and}$$

$$E_2 = 200 - 45 \times 0.4 = 182 \text{ V.}$$

With reference to *Problem 3*, the relationship $\frac{E_1}{E_2} = \frac{\Phi_1 n_1}{\Phi_2 n_2}$ applies to both generators and motors. Since the flux is constant, $\Phi_1 = \Phi_2$.

Hence $\frac{188}{182} = \frac{\Phi_1 \times \frac{1350}{60}}{\Phi_1 \times n_2}$, i.e., $n_2 = \frac{22.5 \times 182}{188} = 21.78$ rev/s.

Thus the speed of the motor when the armature current is 45 A is 21.78×60 rev/min, i.e., **1307 rev/min.**

Problem 5 The shaft torque of a diesel motor driving a 100 V d.c. shunt-wound generator is 25 N m. The armature current of the generator is 16 A at this value of torque. If the shunt field regulator is adjusted so that the flux is reduced by 15%, the torque increases to 35 N m. Determine the armature current at this new value of torque.

The shaft torque T of a generator is proportional to ΦI_a, where Φ is the flux and I_a is the armature current. Thus, $T = k\Phi I_a$, where k is a constant. The torque at flux Φ_1 and armature current I_{a1} is $T_1 = k\Phi_1 I_{a1}$.

Similarly, $T_2 = k\Phi_2 I_{a2}$

By division, $\frac{T_1}{T_2} = \frac{k\Phi_1 I_{a1}}{k\Phi_2 I_{a2}} = \frac{\Phi_1 I_{a1}}{\Phi_2 I_{a2}}$

Hence, $\frac{25}{35} = \frac{\Phi_1 \times 16}{0.85\Phi_1 \times I_{a2}}$

i.e., $I_{a2} = \frac{16 \times 35}{0.85 \times 25} = 26.35$ A.

That is, **the armature current at the new value of torque is 26.35 A.**

Problem 6 A 220 V d.c. shunt-wound motor runs at 800 rev/min and the armature current is 30 A. The armature circuit resistance is 0.4Ω. Determine (a) the maximum value of armature current if the flux is suddenly reduced by 10%; and (b) the steady state value of the armature current at the new value of flux, assuming the shaft torque of the motor remains constant.

(a) For a d.c. shunt-wound motor, $E = V - I_a R_a$. Hence initial generated e.m.f. $E_1 = 220 - 30 \times 0.4 = 208$ V.
The generated e.m.f. is also such that $E \alpha \Phi n$, so at the instant the flux is reduced, the speed has not had time to change, and

$$E = 208 \times \frac{90}{100} = 187.2 \text{ V}$$

Hence, the voltage drop due to the armature resistance is 220 – 187.2, i.e. 32.8 V. The **instantaneous value of the current** is 32.8/0.4, i.e. **82 A**. This increase in current is about three times the initial value and causes an increase in torque, ($T \alpha \Phi I_a$). The motor accelerates because of the larger torque value until steady state conditions are reached.

(b) $T \alpha \Phi I_a$ and since the torque is constant,
$\Phi_1 I_{a1} = \Phi_2 I_{a2}$. The flux Φ is reduced by 10%, hence
$\Phi_2 = 0.9 \Phi_1$.
Thus $\Phi_1 \times 30 = 0.9 \Phi_1 \times I_{a2}$

i.e. the steady state value of armature current, $I_{a2} = \frac{30}{0.9} = 33\frac{1}{3}$ **A**

Problem 7 Sketch the torque/current, speed/current and speed/torque characteristics for a d.c. series-wound motor and with reference to e.m.f. and torque relationships for a d.c. machine, explain their shape.

In a series motor, the armature current flows in the field winding and is equal to the supply current I, (see *Fig 4(b)*).

(i) **The torque/current characteristic**
It was shown in para. 7 that torque $T \alpha \Phi I_a$. Since the armature and field currents are the same current, I, in a series machine, then $T \alpha \Phi I$ over a limited range, before magnetic saturation of the magnetic circuit of the motor is reached, (i.e., the linear portion of the $B - H$ curve for the yoke, poles, air gap, brushes and armature in series). Thus $\Phi \alpha I$ and $T \alpha I^2$. After magnetic saturation, Φ almost becomes a constant and $T \alpha I$. Thus the theoretical torque/current characteristic is as shown in *Fig 6(a)*.

A d.c. motor takes a large current on starting and the characteristic shown in *Fig 6(a)* shows that the series-wound motor has a large torque when the current is large. Hence these motors are used for traction (such as trains, milk delivery vehicles, etc), driving fans and for cranes and hoists, where a large initial torque is required.

(ii) **The speed/current characteristic**
It was shown in para. 10(ii) that

$$n \alpha \frac{V - I_a R_a}{\Phi}$$

In a series motor, $I_a = I$ and below the magnetic saturation level, $\Phi \alpha I$. Thus

$$n \alpha \frac{V - IR}{I}$$

where R is the combined resistance of the series field and armature circuit. Since IR is small compared with V, then an approximate relationship for the speed is $n \alpha \frac{1}{I}$. Hence the theoretical speed/current characteristic is as shown in *Fig 6(b)*. The high speed at small values of current indicate that this type of motor must not be run on very light loads and invariably, such motors are permanently coupled to their loads.

(iii) The theoretical speed/torque characteristic may be derived from (i) and (ii) above by obtaining the torque and speed for various values of current and plotting the co-ordinates on the speed/torque characteristic. A typical speed/torque characteristic is shown in *Fig 6(c)*.

C. FURTHER PROBLEMS ON ELECTRIC MOTOR APPLICATIONS

(a) SHORT ANSWER PROBLEMS

1 An electric . . . . converts electrical energy to mechanical energy.
2 An electric motor converts . . . . energy to . . . . energy.
3 The efficiency of an electrical machine is given by the ratio

$$\frac{\ldots\ldots\ldots\ldots}{\ldots\ldots\ldots\ldots}\ \%.$$

In *Problems 4 to 7*, an electrical machine runs at n rev/s, has a shaft torque of T and takes a current of I from a supply of voltage V.

4 The power input to a generator is . . . . watts.
5 The power input to a motor is . . . . watts.
6 The power output from a generator is . . . . watts.
7 The power output from a motor is . . . . watts.
8 The generated e.m.f. of a d.c. machine is proportional to . . . . volts.
9 The torque produced by a d.c. motor is proportional to . . . . . . N m.
10 In a series-wound d.c. machine, the field winding is in . . . . with the armature circuit.
11 A d.c. motor has its field winding in parallel with the armature circuit. It is called a . . . . wound motor.
12 In a d.c. generator, the relationship between the generated voltage, terminal voltage, current and armature resistance is given by E = ...
13 The equation relating the generated e.m.f., terminal voltage, armature current and armature resistance for a d.c. motor is . . . . = ...
14 Since a d.c. shunt motor is essentially a constant . . . . machine, it may be used to drive . . . .
15 A d.c. series motor has a . . . . initial torque and may be used to drive . . . .
16 A compound-wound d.c. motor has both a . . . . and a . . . . field.
17 The winding on the stator of a three-phase induction motor produces a . . . . magnetic field, which is constant in . . . . and . . . .
18 The winding on the stator of a single-phase induction motor with no special devices produces a . . . . magnetic field.
19 The majority of drives in industry are . . . . motors.
20 A . . . . motor is used to drive an electric clock.

(b) MULTI-CHOICE PROBLEMS (answers on page 163)

1 Which of the following statements is false?

(a) A d.c. motor converts electrical energy to mechanical energy.

(b) The efficiency of a d.c. motor is the ratio $\frac{\text{input power}}{\text{output power}} \times 100\%$.

(c) A d.c. generator converts mechanical energy to electrical energy.

(d) The efficiency of a d.c. generator is the ratio $\frac{\text{output power}}{\text{input power}} \times 100\%$.

A shunt-wound d.c. machine is running at n rev/s and has a shaft torque of T N m. The supply current is I A when connected to d.c. bus bars of voltage V volts. The armature resistance of the machine is R_a ohms, the armature current is I_a A and the generated voltage is E volts. Use this data to find the equations of the quantities stated in *Problems 2 to 9*, selecting the correct answer from the list given below.

(a) $V - I_a R_a$, (b) $E + I_a R_a$; (c) VI;
(d) $E - I_a R_a$; (e) $(T)(2\pi n)$; (f) $V + I_a R_a$.

2 The input power when running as a generator.
3 The output power when running as a motor.
4 The input power when running as a motor.
5 The output power when running as a generator.
6 The generated voltage when running as a motor.
7 The terminal voltage when running as a generator.
8 The generated voltage when running as a generator.
9 The terminal voltage when running as a motor.

10 Which of the following statements is false?
(a) A commutator is necessary as part of a d.c. motor to keep the armature rotating in the same direction.
(b) A commutator is necessary as part of a d.c. generator to produce a uni-directional voltage at the terminals of the generator.
(c) The field winding of a d.c. machine is housed in slots on the armature.
(d) The brushes of a d.c. machine are usually made of carbon and do not rotate with the armature.

11 If the speed of a d.c. machine is doubled and the flux remains constant, the generated e.m.f.
(a) remains the same; (b) is doubled; (c) is halved.

12 If the flux per pole of a shunt-wound d.c. generator is increased, and all other variables are kept the same, the speed
(a) decreases; (b) stays the same; (c) increases.

13 If the flux per pole of a shunt-wound d.c. generator is halved, the generated e.m.f. at constant speed
(a) is doubled; (b) is halved; (c) remains the same.

14 Which of the following statements if false?
(a) A series-wound motor has a large starting torque.
(b) A shunt-wound motor must be permanently connected to its load.
(c) The speed of a series-wound motor drops considerably when load is applied.
(d) A shunt-wound motor is essentially a constant speed machine.

(c) CONVENTIONAL PROBLEMS

1 A 250 V, series-wound motor is running at 500 rev/min and its shaft power is 130 N m. If its efficiency at this load is 88%, find the current taken from the supply.

[30.94 A]

2 In a test on a d.c. motor, the following data was obtained.
Supply voltage: 500 V. Current taken from the supply 42.4 A.
Speed: 850 rev/min. Shaft torque: 187 N m.
Determine tne efficiency of the motor correct to the nearest 0.5%. [78.5%]

3 The shaft torque required to drive a d.c. generator is 18.7 N m when it is running at 1250 rev/min. If its efficiency is 87% under these conditions and the armature current is 17.3 A, determine the voltage at the terminals of the generator. [123.1 V]

4 A 220 V d.c. generator supplies a load of 37.5 A and runs at 1550 rev/min. Determine the shaft torque of the diesel motor driving the generator, if the generator efficiency is 78%. [65.2 N m]

5 Describe the need for a commutator on the armature of a d.c. motor, using diagrams to illustrate your answer.

6 Explain the function of a commutator on the armature of a d.c. motor.

7 Draw a labelled diagram showing a cross-section of a two-pole d.c. machine. Describe the functions performed by the field windings, the armature, the commutator and the brushes.

8 Determine the generated e.m.f. of a d.c. machine if the armature resistance is 0.1Ω and it (a) is running as a motor connected to a 230 V supply, the armature current being 60 A, and (b) is running as a generator with a terminal voltage of 230 V, the armature current being 80 A. [(a) 224 V; (b) 238 V]

9 A d.c. motor has a speed of 900 rev/min when connected to a 460 V supply. Find the approximate value of the speed of the motor when connected to a 200 V supply, assuming the flux decreases by 30% and neglecting the armature volt drop. [559 rev/min]

10 A d.c. generator has a generated e.m.f. of 210 V when running at 700 rev/min and the flux per pole is 120 mWb. Determine the generated e.m.f.
(a) at 1050 rev/min, assuming the flux remains constant;
(b) if the flux is reduced by $\frac{1}{6}$ th at constant speed, and
(c) at a speed of 1155 rev/min and a flux of 132 mWb.
[(a) 315 V; (b) 175 V; (c) 381.2 V]

11 A 250 V d.c. shunt-wound generator has an armature resistance of 0.1Ω. Determine the generated e.m.f. when the generator is supplying 50 kW, neglecting the field current of the generator. [270 V]

12 A series-wound motor is connected to a d.c. supply and develops full-load torque when the current is 30 A and speed is 1000 rev/min. If the flux per pole is proportional to the current flowing, find the current and speed at half full-load torque, when connected to the same supply. [21.2 A, 1415 rev/min]

13 Sketch the theoretical speed/torque characteristic for a series-wound d.c. motor. Use the e.m.f. and torque relationships to explain its shape.

14 Sketch the torque/load characteristic of a series-wound motor and hence give two uses for this type of motor.

15 Using the speed/load characteristic of a shunt-wound machine as the basis of your discussion, give two uses of such a machine.

16 There are three principal types of d.c. motors. Explain the fundamental difference between each of these three types and give two uses for each of the three types.

17 Explain briefly the principal of operation of a three-phase induction motor and state where such motors are used.

18 Three-phase synchronous motors are not in widespread use. By considering the advantage and disadvantages of such motors, explain why this is so.

5 Alternating voltages and currents

A. FORMULAE AND DEFINITIONS ASSOCIATED WITH ALTERNATING VOLTAGES AND CURRENTS

1 Electricity is produced by generators at power stations and then distributed by a vast network of transmission lines (called the National Grid system) to industry and for domestic use. It is easier and cheaper to generate **alternating current** (ac) than direct current (dc) and ac is more conveniently distributed than dc since its voltage can be readily altered using transformers. Whenever dc is needed in preference to ac, devices called rectifiers are used for conversion.

2 Let a single turn coil be free to rotate at constant angular velocity ω symmetrically between the poles of a magnet system as shown in *Fig 1*. An emf is generated in coil (from Faraday's Law) which varies in magnitude and reversed its direction at regular intervals. The reason for this is shown in *Fig 2*.
In positions (a), (e) and (i) the conductors of the loop are effectively moving along the magnetic field, no flux is cut and hence no emf is induced. In position (c) maximum flux is cut and hence maximum emf is induced. In position (g), maximum flux is cut and hence maximum emf is again induced.

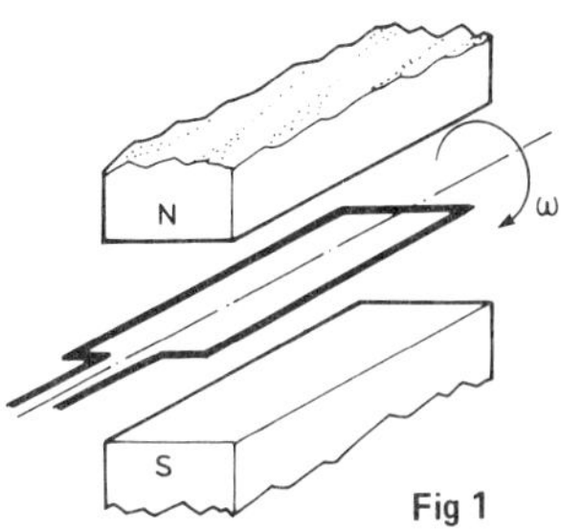

Fig 1

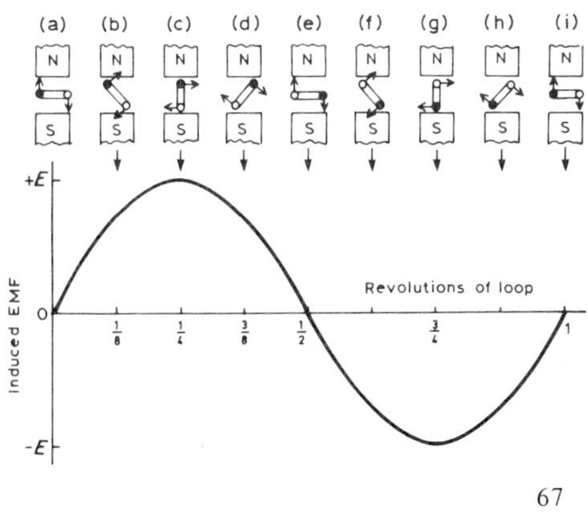

Fig 2

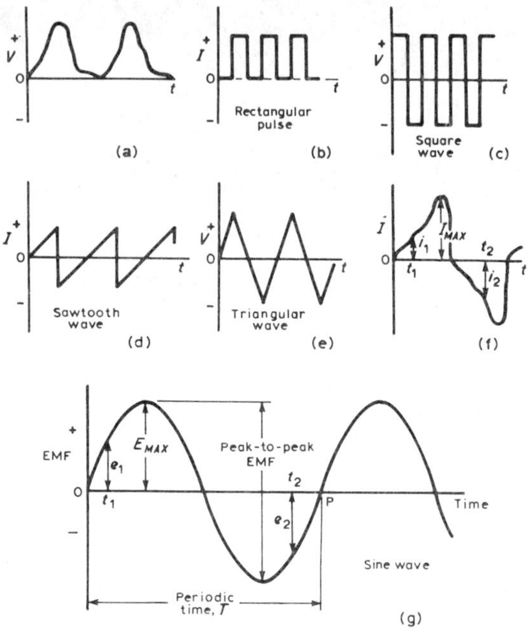

Fig 3

However, using Fleming's right-hand rule, the induced emf is in the opposite direction to that in position (c) and is thus shown as $-E$. In positions (b), (d), (f) and (h) some flux is cut and hence some emf is induced. If all such positions of the coil are considered, in one revolution of the coil, one cycle of alternating emf is produced as shown. This is the principle of operation of the **ac generator** (i.e. the **alternator**).

3 If values of quantities which vary with time t are plotted to a base of time, the resulting graph is called a **waveform**. Some typical waveforms are shown in *Fig 3*. Waveforms (a) and (b) are **unidirectional waveforms**, for, although they vary considerably with time, they flow in one direction only (i.e. they do not cross the time axis and become negative). Waveforms (c) to (g) are called **alternating waveforms** since their quantities are continually changing in direction (i.e. alternately positive and negative).

4 A waveform of the type shown in *Fig 3(g)* is called a **sine wave**. It is the shape of the waveform of emf produced by an alternator and thus the mains electricity supply is of 'sinusoidal' form.

5 One complete series of values is called a **cycle** (i.e. from 0 to P in *Fig 3(g)*).

6 The time taken for an alternating quantity to complete one cycle is called the **period** or the **periodic time**, T, of the waveform.

7 The number of cycles completed in one second is called the **frequency**, f, of the supply and is measured in **hertz**, Hz. The standard frequency of the electricity supply in Great Britain is 50 Hz.

$T = \frac{1}{f}$ or $f = \frac{1}{T}$.

8 **Instantaneous values** are the values of the alternating quantities at any instant of time. They are represented by small letters, i, v, e etc., (see *Figs 3(f)* and *(g)*).

9 The largest value reached in a half cycle is called the **peak value** or the **maximum value** or the **crest value** or the **amplitude** of the waveform. Such values are represented by V_{MAX}, I_{MAX} etc. (see *Figs 3(f)* and *(g)*). A **peak-to-peak** value of emf is shown in *Fig 3(g)* and is the difference between the maximum and minimum values in a cycle.

10 The **average or mean value** of a symmetrical alternating quantity, (such as a sine wave), is the average value measured over a half cycle, (since over a complete cycle the average value is zero).

$$\textbf{Average or mean value} = \frac{\textbf{area under the curve}}{\textbf{length of base}}$$

The area under the curve is found by approximate methods such as the trapezoidal rule, the mid-ordinate rule or Simpson's rule. Average values are represented by V_{AV}, I_{AV}, etc.

For a sine wave, average value = 0.637 × maximum value (i.e. $2/\pi$ × maximum value).

11 The **effective value** of an alternating current is that current which will produce the same heating effect as an equivalent direct current. The effective value is called the **root mean square (rms) value** and whenever an alternating quantity is given, it is assumed to be the rms value. For example, the domestic mains supply in Great Britain is 240 V and is assumed to mean '240 V rms'. The symbols used for rms values are I, V, E, etc. For a non-sinusoidal waveform as shown in *Fig 4* the rms value is given by:

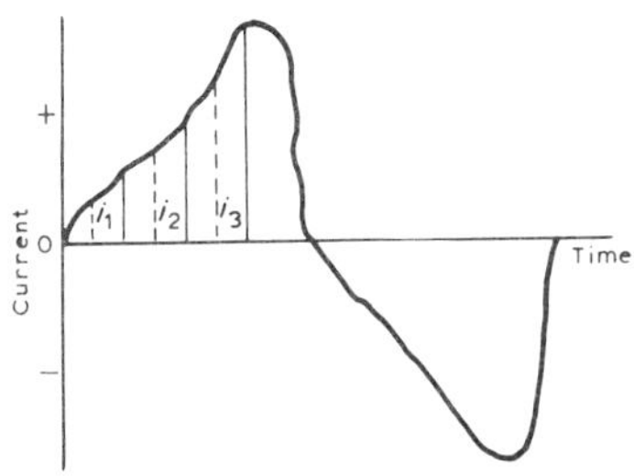

$$I = \sqrt{\left[\frac{i_1^2 + i_2^2 + \ldots + i_n^2}{n}\right]}$$

where n is the number of intervals used.

For a sine wave, rms value = 0.707 × maximum value (i.e. $1/\sqrt{2}$ × maximum value).

Fig 4

12 (a) Form factor $= \dfrac{\text{rms value}}{\text{average value}}$. For a sine wave, form factor = 1.11.

(b) Peak factor $= \dfrac{\text{maximum value}}{\text{rms value}}$. For a sine wave, peak factor = 1.41.

The values of form and peak factors give an indication of the shape of waveforms.

13 In *Fig 5*, OA represents a vector that is free to rotate anticlockwise about 0 at an angular velocity of ω rad/s. A rotating vector is known as a **phasor**. After time t seconds the vector OA has turned through an angle ωt. If the line BC is constructed perpendicular to OA as shown, then

$\sin \omega t = \dfrac{BC}{OB}$ i.e. $BC = OB \sin \omega t$.

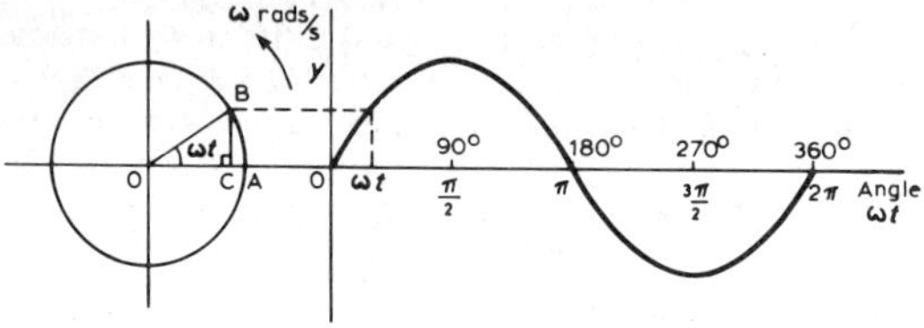

Fig 5

If all such vertical components are projected on to a graph of y against angle ωt (in radians), a sine curve results of maximum value OA. Any quantity which varies sinusoidally can thus be represented as a phasor.

14 A sine curve may not always start at 0°. To show this a periodic function is represented by $y = \sin(\omega t \pm \phi)$, where ϕ is a phase (or angle) difference compared

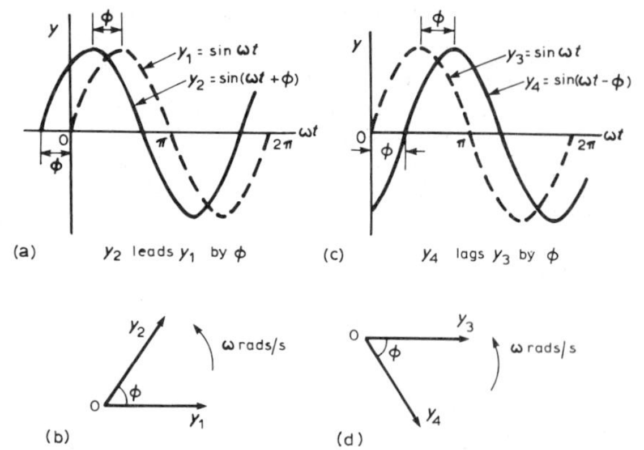

Fig 6

with $y = \sin \omega t$. In *Fig 6(a)*, $y_2 = \sin(\omega t + \phi)$ starts ϕ radians earlier than $y_1 = \sin \omega t$ and is thus said to **lead** y_1 by ϕ radians. Phasors y_1 and y_2 are shown in *Fig 6(b)* at the time when $t = 0$. In *Fig 6(c)*, $y_4 = \sin(\omega t - \phi)$ starts ϕ radians later than $y_3 = \sin \omega t$ and is thus said to **lag** y_3 by ϕ radians. Phasors y_3 and y_4 are shown in *Fig 6(d)* at the time when $t = 0$.

15 Given the general sinusoidal voltage, $v = V_{MAX} \sin(\omega t \pm \phi)$, then
- (i) Amplitude of maximum value = V_{MAX}
- (ii) Peak to peak value = $2V_{MAX}$
- (iii) Angular velocity = ω rads/s
- (iv) Periodic time, $T = 2\pi/\omega$ seconds.
- (v) Frequency, $f = \omega/2\pi$ Hz (hence $\omega = 2\pi f$).
- (vi) ϕ = angle of lag or lead (compared with $v = V_{MAX} \sin \omega t$).

B. WORKED PROBLEMS ON ALTERNATING VOLTAGES AND CURRENTS

(a) FREQUENCY AND PERIODIC TIME

Problem 1 Determine the periodic time for frequencies of (a) 50 Hz and (b) 20 kHz.

(a) Periodic time $T = \frac{1}{f} = \frac{1}{50} = \mathbf{0.02\ s\ or\ 20\ ms}$

(b) Periodic time $T = \frac{1}{f} = \frac{1}{20\ 000} = \mathbf{0.000\ 05\ s\ or\ 50\ \mu s}$

Problem 2 Determine the frequencies for periodic times of (a) 4 ms, (b) 4 μs.

(a) Frequency $f = \frac{1}{T} = \frac{1}{4 \times 10^{-3}} = \frac{1000}{4} = \mathbf{250\ Hz}$

(b) Frequency $f = \frac{1}{T} = \frac{1}{4 \times 10^{-6}} = \frac{1\ 000\ 000}{4} = \mathbf{250\ 000\ Hz}$ or $\mathbf{250\ kHz}$ or $\mathbf{0.25\ MHz}$

Problem 3 An alternating current completes 5 cycles in 8 ms. What is its frequency?

Time for 1 cycle $= \frac{8}{5}$ ms = 1.6 ms = periodic time T.

Frequency $f = \frac{1}{T} = \frac{1}{1.6 \times 10^{-3}} = \frac{1000}{1.6} = \frac{10\ 000}{16} = \mathbf{625\ Hz}$

Further problems on frequency and periodic time may be found in section C(c), Problems 1 to 3, page 78.

(b) AC VALUES OF NON-SINUSOIDAL WAVEFORMS

Problem 4 For the periodic waveforms shown in *Fig 7* determine for each: (i) frequency; (ii) average value over half a cycle; (iii) rms value; (iv) form factor; and (v) peak factor.

(a) Triangular waveform (Fig 7(a))

(i) Time for 1 complete cycle = 20 ms = periodic time, T.

Hence frequency $f = \frac{1}{T} = \frac{1}{20 \times 10^{-3}} = \frac{1000}{20} = \mathbf{50\ Hz}$

(ii) Area under the triangular waveform for a half cycle

$= \frac{1}{2} \times \text{base} \times \text{height}$

$= \frac{1}{2} \times (10 \times 10^{-3}) \times 200$

= 1 volt second.

Average value of waveform

$= \frac{\text{area under curve}}{\text{length of base}}$

$$= \frac{1 \text{ volt second}}{10 \times 10^{-3} \text{ second}}$$

$$= \frac{1000}{10}$$

$$= \mathbf{100\ V}$$

(iii) In *Fig 7(a)*, the first 1/4 cycle is divided into 4 intervals.

Thus rms value

$$= \sqrt{\left[\frac{i_1{}^2 + i_2{}^2 + i_3{}^2 + i_4{}^2}{4}\right]}$$

$$= \sqrt{\left[\frac{25^2 + 75^2 + 125^2 + 175^2}{4}\right]}$$

$$= \mathbf{114.6\ V}$$

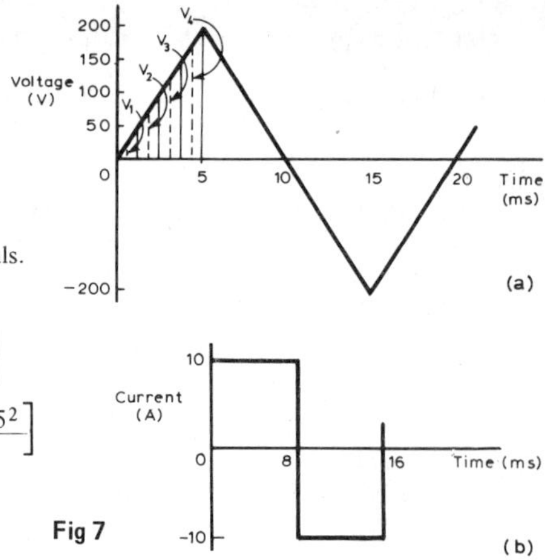

Fig 7

(Note that the greater the number of intervals chosen, the greater the accuracy of the result. For example, if twice the number of ordinates as that chosen above are used, the rms value is found to be 115.6 V)

(iv) Form factor $= \dfrac{\text{rms value}}{\text{average value}} = \dfrac{114.6}{100} = \mathbf{1.15}$

(v) Peak factor $= \dfrac{\text{maximum value}}{\text{rms value}} = \dfrac{200}{114.6} = \mathbf{1.75}$

(b) Rectangular waveform (Fig 7(b))

(i) Time for 1 complete cycle = 16 ms = periodic time, T.

Hence frequency, $f = \dfrac{1}{T} = \dfrac{1}{16 \times 10^{-3}} = \dfrac{1000}{16} = \mathbf{62.5\ Hz}$

(ii) Average value over half a cycle $= \dfrac{\text{area under curve}}{\text{length of base}}$

$$= \frac{10 \times (8 \times 10^{-3})}{8 \times 10^{-3}} = \mathbf{10\ A}$$

(iii) The rms value $= \sqrt{\left[\dfrac{i_1{}^2 + i_2{}^2 + \ldots + i_n{}^2}{n}\right]} = \mathbf{10\ A}$,

however many intervals are chosen, since the waveform is rectangular.

(iv) Form factor $= \dfrac{\text{rms value}}{\text{average value}} = \dfrac{10}{10} = \mathbf{1}$

(v) Peak factor $= \dfrac{\text{maximum value}}{\text{rms value}} = \dfrac{10}{10} = \mathbf{1}$

Problem 5 The following table gives the corresponding values of current and time for a half cycle of alternating current.

time t (ms)	0	0.5	1.0	1.5	2.0	2.5	3.0	3.5	4.0	4.5	5.0
current i (A)	0	7	14	23	40	56	68	76	60	5	0

Assuming the negative half cycle is identical in shape to the positive half cycle, plot the waveform and find (a) the frequency of the supply, (b) the instantaneous values of current after 1.25 ms and 3.8 ms, (c) the peak or maximum value, (d) the mean or average value, and (e) the rms value of the waveform.

The half cycle of alternating current is shown plotted in *Fig 8*.

(a) Time for a half cycle = 5 ms. Hence the time for 1 cycle, i.e. the periodic time, $T = 10$ ms or 0.01 s.

Frequency, $f = \frac{1}{T} = \frac{1}{0.01} =$ **100 Hz**.

(b) Instantaneous value of current after 1.25 ms is **19 A**, from *Fig 8*.
Instantaneous value of current after 3.8 ms is **70 A**, from *Fig 8*.

(c) Peak or maximum value = **76A**

(d) Mean or average value $= \frac{\text{area under curve}}{\text{length of base}}$

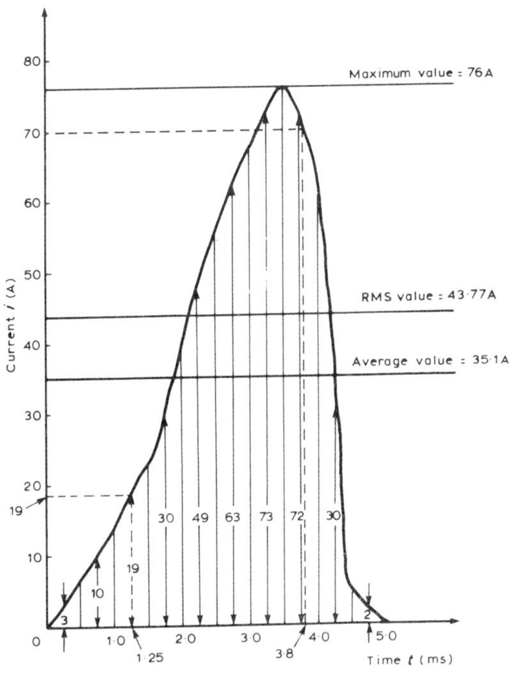

Fig 8

Using the mid-ordinate rule with 10 intervals, each of width 0.5 ms gives:

Area under curve $= (0.5 \times 10^{-3})[3+10+19+30+49+63+73+72+30+2]$ (see *Fig 8*)

$= (0.5 \times 10^{-3})(351)$

Hence mean or average value $= \dfrac{(0.5 \times 10^{-3})(351)}{5 \times 10^{-3}} = \mathbf{35.1\ A}$

(e) rms value $= \sqrt{\left[\dfrac{3^2+10^2+19^2+30^2+49^2+63^2+73^2+72^2+30^2+2^2}{10}\right]}$

$= \sqrt{\left[\dfrac{19\,157}{10}\right]} = \mathbf{43.8\ A}$

Further problems on ac values of non-sinusoidal waveforms may be found in section C(c), Problems 4 to 7, page 78.

(c) AC VALUES OF SINUSOIDAL WAVEFORMS

Problem 6 Calculate the rms value of a sinusoidal current of maximum value 20 A.

For a sine wave,
rms value = 0.707 × maximum value
$= 0.707 \times 20 = \mathbf{14.14\ A}$

Problem 7 Determine the peak and mean values for a 240 V mains supply.

For a sine wave, rms value of voltage $V = 0.707 \times V_{MAX}$
A 240 V mains supply means that 240 V is the rms value.

Hence $V_{MAX} = \dfrac{V}{0.707} = \dfrac{240}{0.707} = \mathbf{339.5\ V = peak\ value}$

Mean value $V_{AV} = 0.637\ V_{MAX} = 0.637 \times 339.5 = \mathbf{216.3\ V}$

Problem 8 A supply voltage has a mean value of 150 V. Determine its maximum value and its rms value.

For a sine wave, mean value = 0.637 × maximum value.

Hence maximum value $= \dfrac{\text{mean value}}{0.637} = \dfrac{150}{0.637} = \mathbf{235.5\ V}$

rms value = 0.707 × maximum value $= 0.707 \times 235.5 = \mathbf{166.5\ V}$

Further problems on ac values of sinusoidal waveforms may be found in section C(c), Problems 8 to 12, page 79.

(d) $v = V_{MAX} \sin(\omega t \pm \phi)$

Problem 9 An alternating voltage is given by $v = 282.8 \sin 314\, t$ volts. Find (a) the rms voltage, (b) the frequency and (c) the instantaneous value of voltage when $t = 4$ ms.

(a) The general expression for an alternating voltage is $v = V_{MAX} \sin(\omega t \pm \phi)$.
Comparing $v = 282.8 \sin 314\, t$ with this general expression gives the peak voltage as 282.8 V
Hence the rms voltage = 0.707 × maximum value = 0.707 × 282.8 = **200 V**

(b) Angular velocity, $\omega = 314$ rads/s i.e. $2\pi f = 314$

Hence frequency, $f = \frac{314}{2\pi}$ = **50 Hz**

(c) When $t = 4$ ms, $v = 282.8 \sin(314 \times 4 \times 10^{-3})$
$= 282.8 \sin(1.256)$

$$1.256 \text{ radians} = \left(1.256 \times \frac{180}{\pi}\right)^{\circ} = 71.96^{\circ} = 71^{\circ}\ 58'$$

Hence $v = 282.8 \sin 71^{\circ}\ 58'$
= **268.9 V**

Problem 10 An alternating voltage is given by $v = 75 \sin(200\pi t - 0.25)$ volts. Find (a) the amplitude; (b) the peak-to-peak value; (c) the rms value, (d) the periodic time; (e) the frequency; and (f) the phase angle (in degrees and minutes) relative to $75 \sin 200\pi t$.

Comparing $v = 75 \sin(200\pi t - 0.25)$ with the general expression $v = V_{MAX} \sin(\omega t \pm \phi)$ gives:

(a) Amplitude, or peak value = **75 V**

(b) Peak-to-peak value = 2 × 75 = **150 V**

(c) The rms value = 0.707 × maximum value = 0.707 × 75 = **53 V**

(d) Angular velocity, $\omega = 200\pi$ rads/s

Hence periodic time, $T = \frac{2\pi}{\omega} = \frac{2\pi}{200\pi} = \frac{1}{100}$ = **0.01 s or 10 ms**

(e) Frequency, $f = \frac{1}{T} = \frac{1}{0.01}$ = **100 Hz**

(f) Phase angle, $\phi = 0.25$ radians lagging $75 \sin 200\pi t$

$$0.25 \text{ rads} = \left(0.25 \times \frac{180}{\pi}\right)^{\circ} = 14.32^{\circ} = 14^{\circ}\ 19'$$

Hence phase angle = **14° 19′ lagging**

Problem 11 An alternating voltage, v, has a periodic time of 0.01 s and a peak value of 40 V. When time t is zero, $v = -20$ V. Express the instantaneous voltage in the form $v = V_{MAX} \sin(\omega t + \phi)$ V.

Amplitude, $V_{MAX} = 40$ V.

Periodic time $T = \frac{2\pi}{\omega}$. Hence angular velocity, $\omega = \frac{2\pi}{T} = \frac{2\pi}{0.01} = 200\pi$ rads/s

$v = V_{MAX} \sin(\omega t + \phi)$ thus becomes $v = 40 \sin(200\pi t + \phi)$ V.

When time $t = 0$, $v = -20$ V

i.e. $-20 = 40 \sin \phi$

$$\sin \phi = \frac{-20}{40} = -0.5$$

Hence $\phi = \arcsin -0.5 = -30° = \left(-30 \times \frac{\pi}{180}\right)$ rads $= -\frac{\pi}{6}$ rads

Thus $v = 40 \sin(200\pi t - \frac{\pi}{6})$ V

Problem 12 The current in an ac circuit at any time t seconds is given by: $i = 120 \sin(100\pi t + 0.36)$ amperes. Find:

(a) the peak value, the periodic time, the frequency and phase angle relative to $120 \sin 100\pi t$;
(b) the value of the current when $t = 0$;
(c) the value of the current when $t = 8$ ms;
(d) the time when the current first reaches 60 A, and
(e) the time when the current is first a maximum.

(a) Peak value = **120 A.**

Periodic time $T = \frac{2\pi}{\omega} = \frac{2\pi}{100\pi}$ (since $\omega = 100\pi$)

$$= \frac{1}{50} = \mathbf{0.02\ s\ or\ 20\ ms}$$

Frequency $f = \frac{1}{T} = \frac{1}{0.02} = $ **50 Hz**

Phase angle = 0.36 rads $= \left(0.36 \times \frac{180}{\pi}\right)^{\circ} = $ **20° 38′ leading**

(b) When $t = 0$, $i = 120 \sin(0+0.36) = 120 \sin 20° 38' = $ **42.29 A**

(c) When $t = 8$ ms, $i = 120 \sin\left[100\pi\left(\frac{8}{10^3}\right) + 0.36\right]$

$= 120 \sin 2.8733$
$= 120 \sin 164° 38'$
$=$ **31.80 A**

(d) When $i = 60$ A, $60 = 120 \sin(100\pi t + 0.36)$

$$\frac{60}{120} = \sin(100\pi t + 0.36)$$

$(100\pi t + 0.36) = \arcsin 0.5 = 30° = \frac{\pi}{6}$ rads $= 0.5236$ rads

Hence time, $t = \frac{0.5236 - 0.36}{100\pi} = $ **0.5208 ms**

(e) When the current is a maximum, $i = 120$ A

Thus $120 = 120 \sin(100\pi t + 0.36)$

$1 = \sin(100\pi t + 0.36)$

$(100\pi t + 0.36) = \arcsin 1 = 90° = \frac{\pi}{2}$ rads = 1.5708 rads

Hence time, $t = \frac{1.5708 - 0.36}{100\pi}$ = **3.854 ms**

Further problems on the general expression $v = V_{MAX} \sin(\omega t \pm \phi)$ may be found in section C(c), Problems 13 to 18, page 80.

C. FURTHER PROBLEMS ON ALTERNATING VOLTAGES AND CURRENTS

(a) SHORT ANSWER PROBLEMS

1 Briefly explain the principle of the simple alternator.
2 What is the difference between an alternating and a unidirectional waveform.
3 What is meant by (a) waveform; (b) cycle.
4 The time to complete one cycle of a waveform is called the . . . .
5 What is frequency? Name its unit.
6 The mains supply voltage has a special shape of waveform called a . . . .
7 Define peak value.
8 What is meant by the rms value?
9 The domestic mains electricity supply voltage in Great Britain is . . . .
10 What is the mean value of a sinusoidal alternating emf which has a maximum value of 100 V?
11 The effective value of a sinusoidal waveform is . . . . X maximum value.
12 What is a phasor quantity?
13 Complete the statement: Form factor = $\frac{\ldots\ldots\ldots\ldots}{\ldots\ldots\ldots\ldots}$, and for a sine wave,

form factor = $\frac{\ldots\ldots\ldots\ldots}{\ldots\ldots\ldots\ldots}$.

14 Complete the statement: Peak factor = $\frac{\ldots\ldots\ldots\ldots}{\ldots\ldots\ldots\ldots}$, and for a sine wave,

peak factor = $\frac{\ldots\ldots\ldots\ldots}{\ldots\ldots\ldots\ldots}$.

15 A sinusoidal current is given by $i = I_{MAX} \sin(\omega t \pm \alpha)$. What do the symbols I_{MAX}, ω and α represent?

(b) MULTI-CHOICE PROBLEMS (answers on page 163)

1 The value of an alternating current at any given instant is (a) a maximum value; (b) a peak value; (c) an instantaneous value; (d) an rms value.
2 An alternating current completes 100 cycles in 0.2 s. Its frequency is (a) 20 Hz; (b) 100 Hz; (c) 0.002 Hz; (d) 1 kHz.
3 In *Fig 9*, at the instant shown the generated emf will be (a) zero; (b) an rms value; (c) an average value; (d) a maximum value.
4 The supply of electrical energy for a consumer is usually by ac because
(a) transmission and distribution are more easily effected;
(b) it is most suitable for variable speed motors;
(c) the volt drop in cables is minimal; (d) cable power losses are negligible.

5 Which of the following statements is false?
(a) It is cheaper to use ac than dc.
(b) Distribution of ac is more convenient than with dc since voltages may be readily altered using transformers.
(c) An alternator is an ac generator
(d) A rectifier changes dc into ac.

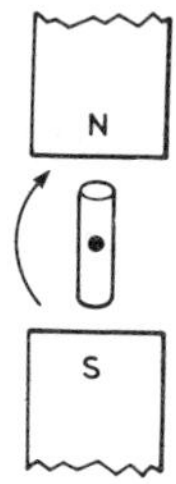

Fig 9

6 An alternating voltage of maximum value 100 V is applied to a lamp. Which of the following direct voltages, if applied to the lamp, would cause the lamp to light with the same brilliance?
(a) 100 V; (b) 63.7 V; (c) 70.7 V; (d) 141.4 V.
7 The value normally stated when referring to alternating currents and voltages is the
(a) instantaneous value; (b) rms value; (c) average value; (d) peak value.
8 State which of the following is false. For a sine wave:
(a) the peak factor is 1.414.
(b) the rms value is 0.707 × peak value.
(c) the average value is 0.637 × rms value.
(d) the form factor is 1.11.
9 An ac supply is 70.7 V, 50 Hz. Which of the following statements is false?
(a) The periodic time is 20 ms.
(b) The peak value of the voltage is 70.7 V.
(c) The rms value of the voltage is 70.7 V.
(d) The peak value of the voltage is 100 V.
10 An alternating voltage is given by $v = 100 \sin (50\pi t - 0.30)$ V. Which of the following statements is true?
(a) The rms voltage is 100 V;
(b) The periodic time is 20 ms;
(c) The frequency is 25 Hz;
(d) The voltage is leading $v = 100 \sin 50\pi t$ by 0.30 radians.

(c) CONVENTIONAL PROBLEMS

Frequency and periodic time

1 Determine the periodic time for the following frequencies:
(a) 2.5 Hz; (b) 100 Hz; (c) 40 kHz.

[(a) 0.4 s; (b) 10 ms; (c) 25 μs]

2 Calculate the frequency for the following periodic times:
(a) 5 ms; (b) 50 μs; (c) 0.2 s.

[(a) 0.2 kHz; (b) 20 kHz; (c) 5 Hz]

3 An alternating current completes 4 cycles in 5 ms. What is its frequency?

[800 Hz]

a.c. values of non-sinusoidal waveforms

4 An alternating current varies with time over half a cycle as follows;

Current (A)	0	0.7	2.0	4.2	8.4	8.2	2.5	1.0	0.4	0.2	0
time (ms)	0	1	2	3	4	5	6	7	8	9	10

The negative half cycle is similar. Plot the curve and determine:
(a) the frequency; (b) the instantaneous values at 3.4 ms and 5.8 ms; (c) its mean value and (d) its rms value.

[(a) 50 Hz. (b) 5.5 A, 3.4 A; (c) 2.8 A; (d) 4.0 A]

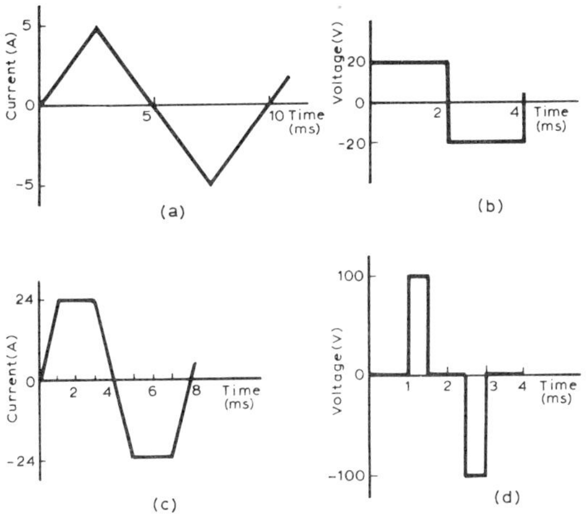

Fig 10

5 For the waveforms shown in *Fig 10* determine for each (i) the frequency; (ii) the average value over half a cycle; (iii) the rms value; (iv) the form factor; (v) the peak factor.

[(a) (i) 100 Hz; (ii) 2.50A; (iii) 2.88A; (iv) 1.15; (v) 1.74
(b) (i) 250 Hz; (ii) 20 V; (iii) 20 V; (iv) 1.0; (v) 1.0
(c) (i) 125 Hz; (ii) 18 A; (iii) 19.56 A;(iv) 1.09; (v) 1.23
(d) (i) 250 Hz; (ii) 25 V; (iii) 50 V; (iv) 2.0; (v) 2.0]

6 An alternating voltage is triangular in shape, rising at a constant rate to a maximum of 300 V in 8 ms and then falling to zero at a constant rate in 4 ms. The negative half cycle is identical in shape to the positive half cycle. Calculate (a) the mean voltage over half a cycle, and (b) the rms voltage.

[(a) 150 V; (b) 171 V]

7 An alternating emf varies with time over half a cycle as follows:

emf (V)	0	45	80	155	215	320	210	95	0
time (ms)	0	1.5	3.0	4.5	6.0	7.5	9.0	10.5	12.0

The negative half cycle is identical in shape to the positive half cycle. Plot the waveform and determine (a) the periodic time and frequency; (b) the instantaneous value of voltage at 3.75 ms, (c) the times when the voltage is 125 V; (d) the mean value, and (e) the rms value.

[(a) 24 ms, 41.67 Hz; (b) 115 V; (c) 4 ms and 10.1 ms; (d) 142 V; (e) 171 V]

a.c. values of sinusoidal waveforms

8 Calculate the rms value of a sinusoidal curve of maximum value 300 V.

[212.1 V]

9 Find the peak and mean values for a 200 V mains supply.

[282.9 V; 180.2 V]

10 Plot a sine wave of peak value 10.0 A. Show that the average value of the waveform is 6.37 A over half a cycle, and that the rms value is 7.07 A.

11 A sinusoidal voltage has a maximum value of 120 V. Calculate its rms and average values.

[84.8 V; 76.4 V]

12 A sinusoidal current has a mean value of 15.0 A. Determine its maximum and rms values.

[23.55 A; 16.65 A]

$v = V_{MAX} \sin(\omega t \pm \phi)$

13 An alternating voltage is represented by $v = 20 \sin 157.1\, t$ volts. Find (a) the maximum value; (b) the frequency; and (c) the periodic time. (d) What is the angular velocity of the phasor representing this waveform?

[(a) 20 V; (b) 25 Hz; (c) 0.04 s; (d) 157.1 rads/s]

14 Find the peak value, the rms value, the periodic time, the frequency and the phase angle (in degrees and minutes) of the following alternating quantities:

(a) $v = 90 \sin 400\pi t$ volts. [(a) 90 V; 63.63 V; 5 ms; 200 Hz; 0°]

(b) $i = 50 \sin(100\pi t + 0.30)$ amperes.

[(b) 50 A; 35.35 A; 0.02 s; 50 Hz; 17°11′ lead]

(c) $e = 200 \sin(628.4t - 0.41)$ volts.

[(c) 200 V; 141.4 V; 0.01 s; 100 Hz; 23°29′ lag]

15 A sinusoidal current has a peak value of 30 A and a frequency of 60 Hz. At time $t = 0$, the current is zero. Express the instantaneous current i in the form $i = I_{MAX} \sin \omega t$.

[$i = 30 \sin 120\pi t$]

16 An alternating voltage v has a periodic time of 20 ms and a maximum value of 200 V. When time $t = 0$, $v = -75$ volts. Deduce a sinusoidal expression for v and sketch one cycle of the voltage showing important points.

[$v = 200 \sin(100\pi t - 0.384)$]

17 The voltage in an alternating current circuit at any time t seconds is given by $v = 60 \sin 40t$ volts. Find the first time when the voltage is (a) 20 V, and (b) −30 V.

[(a) 8.496 ms; (b) 91.63 ms]

18 The instantaneous value of voltage in an ac circuit at any time t seconds is given by $v = 100 \sin(50\pi t - 0.523)$ V. Find:

(a) the peak-to-peak voltage, the periodic time, the frequency and the phase angle;
(b) the voltage when $t = 0$; (c) the voltage when $t = 8$ ms;
(d) the times in the first cycle when the voltage is 60 V;
(e) the times in the first cycle when the voltage is −40 V; and
(f) the first time when the voltage is a maximum.

Sketch the curve for one cycle showing relevant points.

[(a) 200 V, 0.04 s, 25 Hz, 29°58′ lagging; (b) −49.95 V;
(c) 66.96 V; (d) 7.426 ms, 19.23 ms; (e) 25.95 ms, 40.71 ms;
(f) 13.33 ms.]

6 Effects of inductance and capacitance in a.c. circuits

A. MAIN POINTS CONCERNED WITH THE EFFECTS OF INDUCTANCE AND CAPACITANCE IN A.C. CIRCUITS

(a) INDUCTANCE

1 In a purely resistive a.c. circuit, the current I_R and applied voltage V_R are in phase. See *Fig 1(a)*.

2 In a purely inductive a.c. circuit, the current I_L lags the applied voltage V_L by 90° (i.e. $\pi/2$ rads). See *Fig 1(b)*.

3 In a purely inductive circuit the opposition to the flow of alternating current is called the inductive reactance, X_L.

$X_L = \dfrac{V_L}{I_L} = 2\pi f L$ ohms, where f is the supply frequency in hertz and L is the inductance in henry's.

X_L is proportional to f, as shown in *Fig 2*.

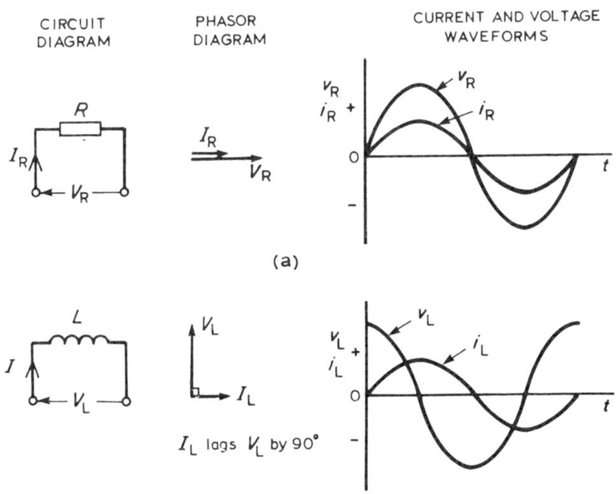

Fig 1

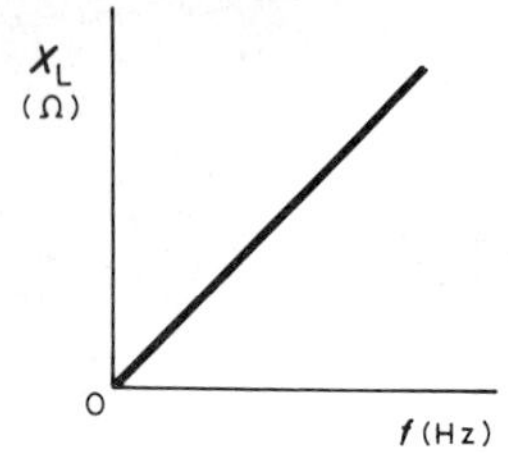

Fig 2

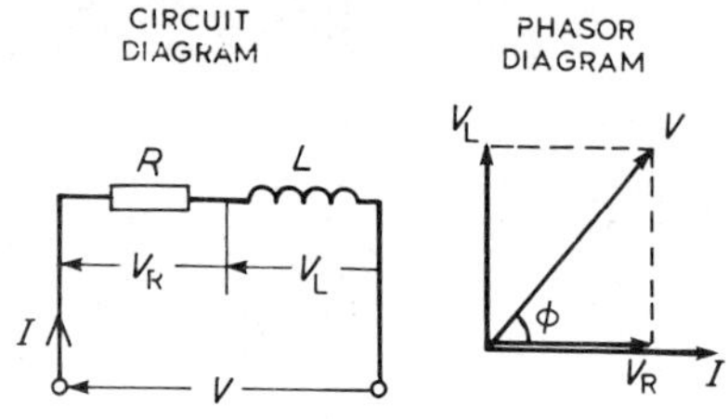

Fig 3

4 In an a.c. series circuit containing inductance L and resistance R, the applied voltage V is the phasor sum of V_R and V_L (see *Fig 3*), and thus the current I lags the applied voltage V by an angle lying between 0° and 90° (depending on the values of V_R and V_L), shown as angle ϕ. In any a.c. series circuit, the current is common to each component and is thus taken as a reference phasor.

5 In an a.c. circuit, the ratio $\dfrac{\text{applied voltage } V}{\text{current } I}$ is called the impedance Z,

i.e. $Z = \dfrac{V}{I}$ ohms.

6 From the phasor diagram of *Fig 3*, the **voltage triangle** is derived, as shown in *Fig 4(a)*.
For the R–L circuit: $V = \sqrt{(V_R^2 + V_L^2)}$ (by Pythagoras' theorem)

and $\tan\phi = \dfrac{V_L}{V_R}$ (by trigonometric ratios)

7 If each side of the voltage triangle in *Fig 4(a)* is divided by current I then the **impedance triangle** is derived, as shown in *Fig 4(b)*.
For the R–L circuit: $Z = \sqrt{(R^2 + X_L^2)}$

$\tan\phi = \dfrac{X_L}{R}$, $\sin\phi = \dfrac{X_L}{Z}$ and $\cos\phi = \dfrac{R}{Z}$

(See *Problems 1 to 7*).

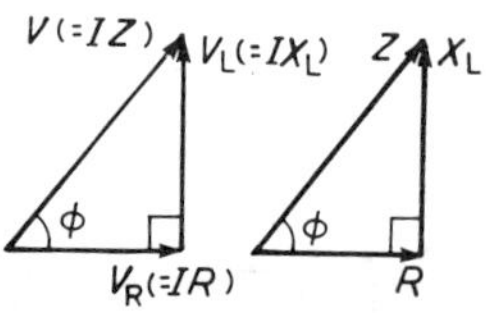

Fig 4

(a) VOLTAGE TRIANGLE (b) IMPEDANCE TRIANGLE

(b) CAPACITANCE

8 In a purely capacitive a.c. circuit, the current I_C leads the applied voltage V_C by 90° (i.e., $\pi/2$ rads). See *Fig 5*.

9 In a purely capacitive circuit the opposition to the flow of alternating current is called the capacitive reactance, X_C.

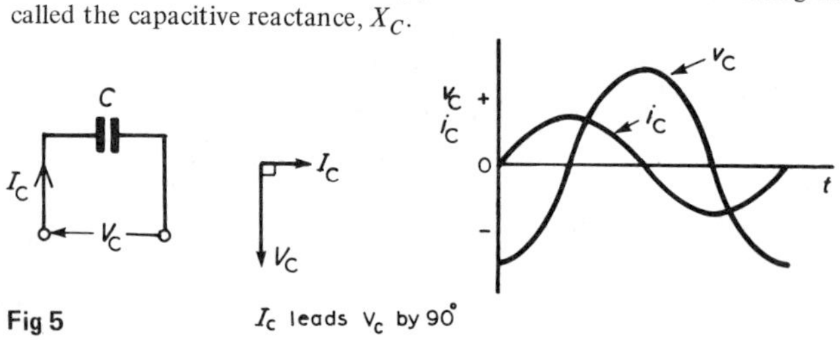

Fig 5

I_C leads V_C by 90°

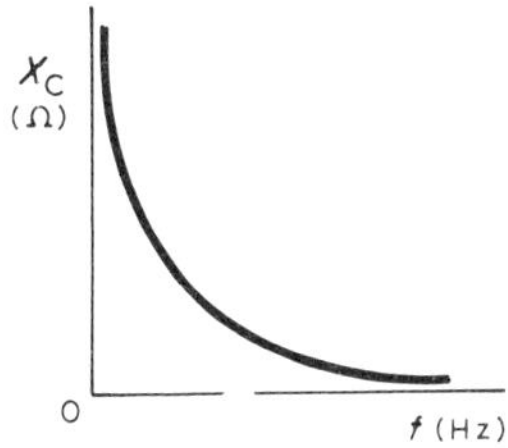

Fig 6

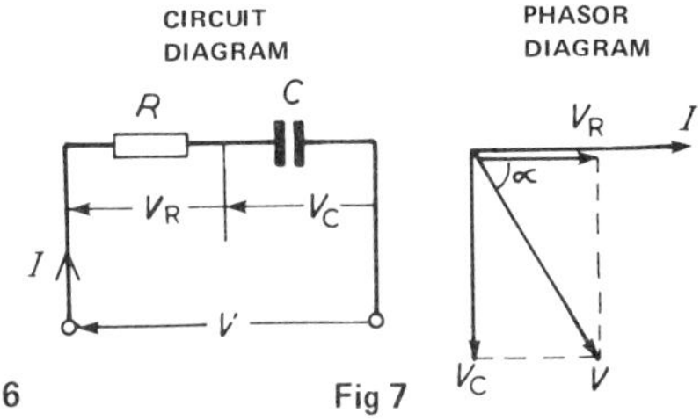

Fig 7

$X_C = \dfrac{V_C}{I_C} = \dfrac{1}{2\pi f C}$ ohms, where C is the capacitance in farads. X_C varies with f as shown in *Fig 6*.

10 In an a.c. series circuit containing capacitance C and resistance R, the applied voltage V is the phasor sum of V_R and V_C (see *Fig 7*) and thus the current I leads the applied voltage V by an angle lying between 0° and 90°, (depending on the values of V_R and V_C), shown as angle α.

11 From the phasor diagram of *Fig 7*, the voltage triangle is derived as shown in *Fig 8(a)*.

For the R–C circuit: $V = \sqrt{(V_R^2 + V_C^2)}$

and $\tan\alpha = \dfrac{V_C}{V_R}$

12 If each side of the voltage triangle in *Fig 8(a)* divided by current I then the impedance triangle is derived, as shown in *Fig 8(b)*.

For the R–C circuit: $Z = \sqrt{(R^2 + X_C^2)}$

$\tan\alpha = \dfrac{X_C}{R}$, $\sin\alpha = \dfrac{X_C}{Z}$ and $\cos\alpha = \dfrac{R}{Z}$

(See *Problems 8 to 12*).

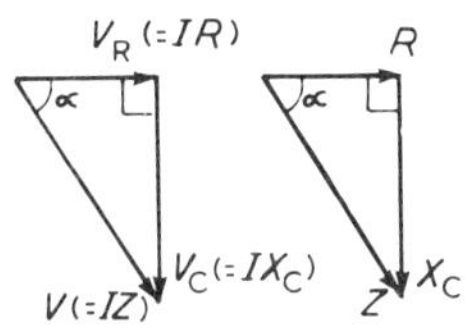

(a) VOLTAGE TRIANGLE (b) IMPEDANCE TRIANGLE

Fig 8

B. WORKED PROBLEMS ON THE EFFECTS OF INDUCTANCE AND CAPACITANCE IN A.C. CIRCUITS

(a) INDUCTANCE

Problem 1 (a) Calculate the reactance of a coil of inductance 0.32 H when it is connected to a 50 Hz supply.
(b) A coil has a reactance of 124 ohms in a circuit with a supply of frequency 5 kHz. Determine the inductance of the coil.

(a) Inductive reactance, $X_L = 2\pi f L = 2\pi(50)(0.32) = \mathbf{100.5\ \Omega}$

(b) Since $X_L = 2\pi f L$, inductance $L = \dfrac{X_L}{2\pi f} = \dfrac{124}{2\pi(5000)}$ H $=$ **3.95 mH.**

Problem 2 A coil has an inductance of 40 mH and negligible resistance. Calculate its inductive reactance and the resulting current if connected to (a) a 240 V, 50 HZ supply, (b) a 100 V, 1 kHz supply.

(a) Inductive reactance $X_L = 2\pi fL = 2\pi(50)(40 \times 10^{-3}) = \mathbf{12.57\ \Omega}$.

$$\text{Current } I = \frac{V}{X_L} = \frac{240}{12.57} = \mathbf{19.09\ A}$$

(b) Inductive reactance $X_L = 2\pi(1000)(40 \times 10^{-3}) = \mathbf{251.3\ \Omega}$

$$\text{Current } I = \frac{V}{X_L} = \frac{100}{251.3} = \mathbf{0.398\ A}$$

Problem 3 In a series R–L circuit, the p.d. across the resistance R is 12 V and the p.d. across the inductance L is 5 V. Find the supply voltage and the phase angle between current and voltage.

From the voltage triangle of *Fig 4(a)*, supply voltage $V = \sqrt{(12^2 + 5^2)}$
$= \mathbf{13\ volts}$

(Note that in a.c. circuits, the supply voltage is **not** the arithmetic sum of the p.d's across components. It is, in fact, the **phasor sum.**)

$\tan \phi = \frac{V_L}{V_R} = \frac{5}{12}$, from which $\phi = \arctan \frac{5}{12} = \mathbf{22° \ 37' \ lagging.}$

('Lagging' infers that the current is 'behind' the voltage, since phasors revolve anticlockwise.)

Problem 4 A coil has a resistance of 12 ohms and an inductance of 15.9 mH. Calculate (a) the reactance, (b) the impedance, and (c) the current taken from a 240 V, 50 Hz supply. Determine also the phase angle between the supply voltage and current.

$R = 12\Omega$; $L = 15.9$ mH $= 15.9 \times 10^{-3}$ H; $f = 50$ Hz; $V = 240$ V.

(a) Inductive reactance $X_L = 2\pi fL = 2\pi\,(50)(15.9 \times 10^{-3}) = \mathbf{5\Omega}$

(b) Impedance $Z = \sqrt{(R^2 + X_L{}^2)} = \sqrt{(12^2 + 5^2)} = \mathbf{13\Omega}$

(c) Current $I = \frac{V}{Z} = \frac{240}{13} = \mathbf{18.5\ A}$

The circuit and phasor diagrams and the voltage and impedance triangles are as shown in *Figs 3 and 4.*

Since $\tan \phi = \frac{X_L}{R}$, $\phi = \arctan \frac{X_L}{R} = \arctan \frac{5}{12} = \mathbf{22° \ 37' \ lagging.}$

Problem 5 A coil takes a current of 2 A from a 12 V d.c. supply. When connected to a 240 V, 50 Hz a.c. supply the current is 20 A. Calculate the resistance, impedance, inductive reactance and inductance of the coil.

$$\text{Resistance } R = \frac{\text{d.c. voltage}}{\text{d.c. current}} = \frac{12}{2} = 6\Omega$$

$$\text{Impedance } Z = \frac{\text{a.c. voltage}}{\text{a.c. current}} = \frac{240}{20} = 12\Omega$$

As $Z = \sqrt{(R^2 + X_L{}^2)}$, inductive reactance, $X_L = \sqrt{(Z^2 - R^2)}$
$= \sqrt{(12^2 - 6^2)} = 10.39\Omega$

As $X_L = 2\pi fL$, inductance $L = \frac{X_L}{2\pi f} = \frac{10.39}{2\pi(50)} = \mathbf{33.1\ mH}$

This problem indicates a simple method for finding the inductance of a coil, i.e. firstly to measure the current when the coil is connected to a d.c. supply of known voltage, and then to repeat the process with an a.c. supply.

Problem 6 A coil of inductance 318.3 mH and negligible resistance is connected in series with a 200Ω resistor to a 240 V, 50 Hz supply. Calculate (a) the inductive reactance of the coil, (b) the impedance of the circuit, (c) the current in the circuit, (d) the p.d. across each component, and (e) the circuit phase angle.

$L = 318.3$ mH $= 0.3183$ H; $R = 200\Omega$; $V = 240$ V; $f = 50$ Hz.
The circuit diagram is as shown in *Fig 3*.
(a) Inductive reactance $X_L = 2\pi fL = 2\pi(50)(0.3183) =$ **100Ω**
(b) Impedance $Z = \sqrt{(R^2 + X_L^2)} = \sqrt{[(200)^2 + (100)^2]} =$ **223.6Ω**
(c) Current $I = \dfrac{V}{Z} = \dfrac{240}{223.6} =$ **1.073 A.**
(d) The p.d. across the coil, $V_L = I X_L = 1.073 \times 100 =$ **107.3 V**
The p.d. across the resistor, $V_R = I R = 1.073 \times 200 =$ **214.6 V**
[Check: $\sqrt{(V_R^2 + V_L^2)} = \sqrt{[(214.6)^2 + (107.3)^2]} = 240$ V, the supply voltage]
(e) From the impedance triangle, angle $\phi = \arctan \dfrac{X_L}{R} = \arctan \dfrac{100}{200}$

Hence **phase angle $\Phi = 26° 34'$ lagging**

Problem 7 A coil has a resistance of 100 Ω and an inductance of 200 mH. If an alternating voltage v, given by $v = 200 \sin 500\,t$ volts is applied across the coil, calculate (a) the circuit impedance, (b) the current flowing, (c) the p.d. across the resistance, (d) the p.d. across the inductance and (e) the phase angle between voltage and current.

Because $v = 200 \sin 500\,t$ volts, $V_{MAX} = 200$ V and $\omega = 2\pi f = 500$ rad/s.
Hence r.m.s. voltage $V = 0.707 \times 200 = 141.4$ V.
Inductive reactance $X_L = 2\pi fL = \omega L = 500 \times 200 \times 10^{-3} = 100\Omega$
(a) Impedance $Z = \sqrt{(R^2 + X_L^2)} = \sqrt{(100^2 + 100^2)} =$ **141.4Ω**
(b) Current $I = \dfrac{V}{Z} = \dfrac{141.4}{141.4} =$ **1A.**
(c) The p.d. across the resistance $V_R = IR = 1 \times 100 =$ **100 V**
(d) P.D. across the inductance, $V_L = I X_L = 1 \times 100 =$ **100 V.**
(e) Phase angle between voltage and current, $\phi = \arctan \dfrac{X_L}{R} = \arctan \dfrac{100}{100}$

Hence $\phi = 45°$ or $\dfrac{\pi}{4}$ rads

Further problems on the effects of inductance in a.c. circuits may be found in section C(c), Problems 1 to 11, page 88.

(b) CAPACITANCE

Problem 8 Determine the capacitive reactance of a capacitor of 10 μF when connected to a circuit of frequency (a) 50 Hz, (b) 20 kHz.

(a) Capacitive reactance $X_C = \dfrac{1}{2\pi fC} = \dfrac{1}{2\pi(50)(10 \times 10^{-6})} = \dfrac{10^6}{2\pi(50)(10)}$
$=$ **318.3Ω**

(b) $X_C = \dfrac{1}{2\pi fC} = \dfrac{1}{2\pi(20 \times 10^3)(10 \times 10^{-6})} = \dfrac{10^6}{2\pi(20 \times 10^3)(10)} =$ **0.796Ω**

Hence as the frequency is increased from 50 Hz to 20 kHz, X_C decreases from 318.3Ω to 0.796Ω (see *Fig 6*).

Problem 9 A capacitor has a reactance of 40Ω when operated on a 50 Hz supply. Determine the value of its capacitance.

$$X_C = \frac{1}{2\pi fC}, \text{ so capacitance } C = \frac{1}{2\pi fX_C} = \frac{1}{2\pi(50)(40)} \text{ F}$$
$$= \frac{10^6}{2\pi(50)(40)} \mu\text{F} = \mathbf{79.58\mu F}$$

Problem 10 Calculate the current taken by a 23 μF capacitor when connected to a 240 V, 50 Hz supply.

$$\text{Current } I = \frac{V}{X_C} = \frac{V}{\frac{1}{2\pi fC}} = 2\pi fCV = 2\pi(50)(23 \times 10^{-6})(240)$$
$$= \mathbf{1.73\ A}$$

Problem 11 A resistor of 25Ω is connected in series with a capacitor of 45μF. Calculate (a) the impedance, and (b) the current taken from a 240 V, 50 Hz supply. Find also the phase angle between the supply voltage and the current.

$R = 25\Omega$; $C = 45\mu F = 45 \times 10^{-6}$ F; $V = 240$ V; $f = 50$ Hz.
The circuit diagram is as shown in *Fig 7*.

$$\text{Capacitive reactance } X_C = \frac{1}{2\pi fC} = \frac{1}{2\pi(50)(45 \times 10^{-6})} = \mathbf{70.74\Omega}$$

(a) Impedance $Z = \sqrt{(R^2 + X_C^2)} = \sqrt{[(25)^2 + (70.74)^2]} = \mathbf{75.03\Omega}$

(b) Current $I = \frac{V}{Z} = \frac{240}{75.03} = \mathbf{3.20\ A}$

Phase angle between the supply voltage and current, $\alpha = \arctan \frac{X_C}{R}$

Hence $\alpha = \arctan \frac{70.74}{25} = \mathbf{70^\circ\ 32'\ leading.}$

('Leading' infers that the current is 'ahead' of the voltage, as phasors revolve anticlockwise.)

Problem 12 A capacitor *C* is connected in series with a 40Ω resistor across a supply of frequency 60 Hz. A current of 3 A flows and the circuit impedance is 50Ω. Calculate (a) the value of capacitance *C*, (b) the supply voltage, (c) the phase angle between the supply voltage and current, (d) the p.d. across the resistor, and (e) the p.d. across the capacitor. Draw the phasor diagram.

(a) Impedance $Z = \sqrt{(R^2 + X_C^2)}$

Hence $X_C = \sqrt{(Z^2 - R^2)} = \sqrt{(50^2 - 40^2)} = 30\Omega$

$X_C = \frac{1}{2\pi fC}$. Hence $C = \frac{1}{2\pi fX_C} = \frac{1}{2\pi(60)30}$ F $= \mathbf{88.42\ \mu F}$

(b) Since $Z = \frac{V}{I}$ then $V = IZ = (3)(50) = \mathbf{150\ V}$

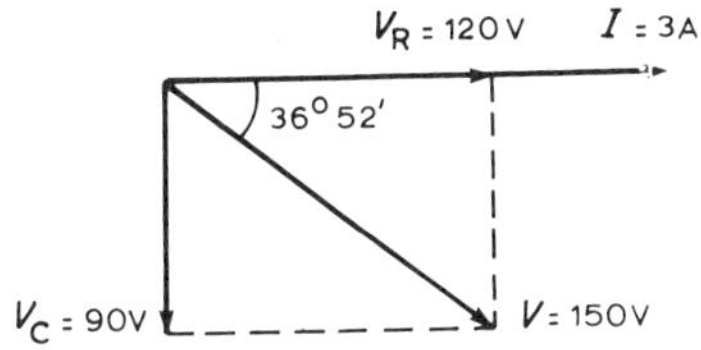

Phasor diagram

Fig 9

(c) Phase angle $\alpha = \arctan \dfrac{X_C}{R} = \arctan \dfrac{30}{40} = \mathbf{36° \ 52' \ leading}$

(d) p.d. across resistor $V_R = IR = (3)(40) = \mathbf{120\ V}$

(e) p.d. across capacitor, $V_C = I X_C = (3)(30) = \mathbf{90\ V}$

The phasor diagram is shown in *Fig 9*.

Further problems on the effects of capacitance in a.c. circuits may be found in section C(c), Problems 12 to 20, page 89.

C. FURTHER PROBLEMS ON THE EFFECTS OF INDUCTANCE AND CAPACITANCE IN A.C. CIRCUITS

(a) SHORT ANSWER PROBLEMS

1 Complete the following statements:
(a) In a purely resistive a.c. circuit the current is . . . . . . with the voltage.
(b) In a purely inductive a.c. circuit the current . . . . the voltage by . . . . . degrees.
(c) In a purely capacitive a.c. circuit the current . . . . . . the voltage by . . . . . degrees.

2 Draw phasor diagrams to represent (a) a purely resistive a.c. circuit, (b) a purely inductive a.c. circuit, and (c) a purely capacitive a.c. circuit.

3 What is inductive reactance? State the symbol and formula for determining inductive reactance.

4 What is capacitive reactance? State the symbol and formula for determining capacitive reactance.

5 What does 'impedance' mean when referring to an a.c. circuit?

6 Draw an impedance triangle for an R–L circuit. Derive from the triangle an expression for (a) impedance (b) phase angle.

(b) MULTI-CHOICE PROBLEMS (answers on page 163)

1 An inductance of 10 mH connected across a 100 V, 50 Hz supply has an inductive reactance of (a) $10\pi\Omega$ (b) $1000\pi\Omega$ (c) $\pi\Omega$ (d) πH.

2 When the frequency of an a.c. circuit containing resistance and inductance is increased, the current (a) decreases, (b) increases, (c) stays the same.

3 In problem 2, the phase angle of the circuit, (a) decreases, (b) increases, (c) stays the same.

4 A capacitor of 1 μF is connected to a 50 Hz supply. The capactive reactance is (a) 50 MΩ (b) $\frac{10}{\pi}$ kΩ (c) $\frac{\pi}{10^4}$ Ω (d) $\frac{10}{\pi}$ Ω

5 In a series a.c. circuit the voltage across a pure inductance is 12 V and the voltage across a pure resistance is 5 V. The supply voltage is (a) 13 V, (b) 17 V, (c) 7 V, (d) 2.4 V.

6 Inductive reactance results in a current that
(a) leads the voltage by 90°
(b) is in phase with the voltage,
(c) leads the voltage by π rads,
(d) lags the voltage by $\frac{\pi}{2}$ rads.

7 The impedance of a coil, which has a resistance of X ohms and an inductance of Y henry's, connected across a supply of frequency K Hz is: (a) $2\pi KY$ (b) $X + Y$ (c) $\sqrt{(X^2 + Y^2)}$ (d) $\sqrt{X^2 + (2\pi KY)^2}$

8 In problem 7, the phase angle between the current and the applied voltage is given by: (a) arctan $\frac{Y}{X}$ (b) arctan $\frac{2\pi KY}{X}$ (c) arctan $\frac{X}{2\pi KY}$ (d) tan $\frac{2\pi KY}{X}$

9 When a capacitor is connected to an a.c. supply the current (a) leads the voltage by 180°, (b) is in phase with the voltage, (c) leads the voltage by $\pi/2$ rads, (d) lags the voltage by 90°.

10 When the frequency of an a.c. circuit containing resistance and capacitance is decreased the current (a) increases, (b) decreases, (c) stays the same.

(c) CONVENTIONAL PROBLEMS

Inductance

1 Calculate the reactance of a coil of inductance 0.2 H when it is connected to (a) a 50 Hz, (b) a 600 Hz (c) a 40 kHz supply.
[(a) 62.83Ω (b) 754Ω (c) 50.27 kΩ]

2 A coil has a reactance of 120Ω in a circuit with a supply frequency of 4 kHz. Calculate the inductance of the coil.
[4.77 mH]

3 A supply of 240 V, 50 Hz is connected across a pure inductance and the resulting current is 1.2 A. Calculate the inductance of the coil.
[0.637 H]

4 An e.m.f. of 200 V at a frequency of 2 kHz is applied to a coil of pure inductance 50 mH. Determine (a) the reactance of the coil, (b) the current flowing in the coil.
[(a) 628Ω (b) 0.318 A]

5 A 120 mH inductor has a 50 mA, 1 kHz alternating current flowing through it. Find the p.d. across the inductor.
[37.7 V]

6 Determine the impedance of a coil which has a resistance of 12Ω and a reactance of 16Ω.
[20Ω]

7 A coil of inductance 80 mH and resistance 60Ω is connected to a 200 V, 100 Hz supply. Calculate the circuit impedance and the current taken from the supply. Find also the phase angle between the current and the supply voltage.
[78.27Ω; 2.555 A; 39° 57′ lagging]

8 An alternating voltage given by $v = 100 \sin 240\,t$ volts is applied across a coil of resistance 32Ω and inductance 100 mH. Determine (a) the circuit impedance, (b) the current flowing, (c) the p.d. across the resistance, (d) the p.d. across the inductance. [(a) 40Ω (b) 1.77 A (c) 56.64 V (d) 42.48 V]

9 A coil takes a current of 5 A from a 20 V d.c. supply. When connected to a 200 V, 50 Hz a.c. supply the current is 25 A. Calculate the (a) resistance, (b) impedance and (c) inductance of the coil. [(a) 4Ω (b) 8Ω (c) 22.06 mH]

10 A resistor and an inductor of negligible resistance are connected in series to an a.c. supply. The p.d. across the resistor is 18 V and the p.d. across the inductor is 24 V. Calculate the supply voltage and the phase angle between voltage and current. [30 V, 53° 8′ lagging]

11 A coil of inductance 636.6 mH and negligible resistance is connected in series with a 100Ω resistor to a 240 V, 50 Hz supply. Calculate (a) the inductive reactance of the coil, (b) the impedance of the circuit, (c) the current in the circuit, (d) the p.d. across each component, and (e) the circuit phase angle.
[(a) 200Ω (b) 223.6Ω (c) 1.118 A (d) 223.6 V, 111.8 V (e) 63° 26′ lagging]

Capacitance

12 Calculate the capacitive reactance of a capacitor of 20 μF when connected to an a.c. circuit of frequency (a) 20 Hz, (b) 50 Hz, (c) 4 kHz.
[(a) 397.9Ω (b) 15.92Ω (c) 1.989Ω]

13 A capacitor has a reactance of 80Ω when connected to a 50 Hz supply. Calculate the value of its capacitance.
[39.79μF]

14 Calculate the current taken by a 10μF capacitor when connected to a 200 V, 100 Hz supply.
[1.257 A]

15 A capacitor has a capacitive reactance of 400Ω when connected to a 100 V, 25 Hz supply. Determine its capacitance and the current taken from the supply.
[15.92μF; 0.25 A]

16 A voltage of 35 V is applied across a *C–R* series circuit. If the voltage across the resistor is 21 V, find the voltage across the capacitor.
[28 V]

17 A resistance of 50Ω is connected in series with a capacitance of 20μF. If a supply of 200 V, 100 Hz is connected across the arrangement, find (a) the circuit impedance, (b) the current flowing, (c) the phase angle between voltage and current.
[(a) 93.98Ω (b) 2.128 A (c) 57° 51′ leading]

18 A 24.87μF capacitor and a 30Ω resistor are connected in series across a 150 V supply. If the current flowing is 3 A find (a) the frequency of the supply, (b) the p.d. across the resistor and (c) the p.d. across the capacitor.
[(a) 160 Hz; (b) 90 V; (c) 120 V]

19 An alternating voltage $v = 250 \sin 800\,t$ volts is applied across a series circuit containing a 30Ω resistor and 50μF capacitor. Calculate (a) the circuit impedance, (b) the current flowing, (c) the p.d. across the resistor, (d) the p.d. across the capacitor, (e) the phase angle between voltage and current.
[(a) 39.05Ω; (b) 4.526 A; (c) 135.8 V; (d) 113.2 V; (e) 39° 48′]

20 A 400Ω resistor is connected in series with a 2358 pF capacitor across a 12 V a.c. supply. Determine the supply frequency if the current flowing in the circuit is 24 mA.
[225 kHz]

7 Arithmetic calculations

A. MAIN POINTS INVOLVED WITH ARITHMETIC CALCULATIONS

1 (i) In all problems in which the measurement of distance, time, mass or other quantities occurs, an exact answer cannot be given: only an answer which is correct to a stated degree of accuracy. This results in what is called an **error due to measurement.**

(ii) To take account of measurement errors it is usual to limit answers so that the result given is **not more than one significant figure greater than the least accurate number given in the data.**

(iii) **Rounding-off errors** can exist with decimal fractions. For example, to state that $\pi = 3.142$ is not strictly correct, but '$\pi = 3.142$ correct to 4 significant figures' is a true statement.

(Actually, $\pi = 3.14159265 \ldots\ldots$)

(iv) It is possible, through an incorrect procedure, to obtain the wrong answer to a calculation. This type of error is known as a **blunder.**

(v) An **order of magnitude error** is said to exist if incorrect positioning of the decimal point occurs after a calculation has been completed.

(vi) Blunders and order of magnitude errors can be reduced by determining **approximate values of calculations.** Answers which do not seem feasible must be checked and the calculation must be repeated as necessary. (See *Problems 1 and 2*).

2 Aids to arithmetic calculations include:

(i) 4 figure tables of common logarithms and antilogarithms;

(ii) slide-rules, and

(iii) calculators.

3 **Common logarithms** may be used when multiplying or dividing numbers or for raising numbers to powers.

(i) **Definition of a logarithm**:

'The common logarithm of a number is the power to which 10 has to be raised to be equal to the number.'

Thus, since $100 = 10^2$, then 2 is the logarithm of 100 to a base 10 and is written as $2 = \log_{10} 100$, or more usually, $2 = \lg 100$ (where lg means $\log_{10}$).

(ii) **Tables of logarithms**

A logarithm is made up of two parts: (a) an integer called the **characteristic,** and (b) a decimal part called the **mantissa** (which is always positive). To find the logarithm of any positive number, initially change the number of standard form. The power to which the 10 is raised gives the characteristic and the mantissa is read directly from logarithm tables, the 4 digit number being written as a number between 0 and 0.9999. For example,

$\lg 3.564 \quad = \lg (3.564 \times 10^0) \quad = 0.5519$

$\lg 356.4 \quad = \lg (3.564 \times 10^2) \quad = 2.5519$

$\lg 0.003564 = \lg (3.564 \times 10^{-3}) \quad = -3 + 0.5519,$

which is written as $\bar{3}.5519$.

(*Note.* $-3 + 0.5519 = -2.4481$ which is the form of the result a calculator gives for lg 0.003564. Thus -2.4481 has the same value as $\bar{3}.5519$.)

Similarly, $\lg 8672 = 3.9381$, $\lg 0.127 = \bar{1}.1038$ and $\lg 1 = \lg 10^0 = 0$.

(iii) **Tables of antilogarithms**

The reverse process to finding the logarithm of a number is to find the antilogarithm of a number which can be determined by using tables. Only the mantissa is used with antilogarithm tables; the characteristic merely fixes the position of the decimal point.

For example, to find the antilogarithm of 3.9381, .9381 in antilogarithmic tables corresponds to 8.672 (note that anti-logarithm tables always give a four-figure number which is then changed into a number between 1.0 and 9.999). The characteristic 3 indicates that 8.672 is multiplied by 10^3. Hence the antilogarithm of 3.9381 is 8672.

Similarly, if $\lg N = 1.7214$, $N = 5.265 \times 10^1 = 52.65$,

and if $\lg N = \bar{2}.3689$, $\quad N = 2.338 \times 10^{-2} = 0.02338$

(iv) **Rules of logarithms**

(a) To multiply two numbers, $\lg (A \times B) = \lg A + \lg B$

(b) To divide two numbers, $\lg\left(\frac{A}{B}\right) = \lg A - \lg B$

(c) To raise a number to a power, $\lg A^n = n \times \lg A$.

(*see Problems 13 to 16*)

4 A **slide rule** provides a method of performing calculations quickly, correct to about 3 significant figure accuracy. However, their use, together with the use of tables, has declined with the advent of inexpensive **electronic calculators**.

The scientific notation type calculator can perform all the functions provided by tables and slide rules more quickly and with greater accuracy (8 digit displays (or more) are available on most calculators). However answers should be limited to a reasonable number of significant figures (see para l(ii)).

There are many varying makes of calculators and slide rules and each usually have detailed instructions for their use supplied at purchase. 4-figure tables should not, however, be completely abandoned since they remain the cheapest of all aids to calculations and are not dependant upon a source of electrical power. Use a calculator and/or slide rule to evaluate the following worked problems.

B. WORKED PROBLEMS INVOLVING ARITHMETIC CALCULATIONS

Problem 1 The area A of a triangle is given by $A = \frac{1}{2}bh$. The base b when measured is found to be 3.26 cm, and the perpendicular height h is 7.5 cm. Determine the area of the triangle.

Area of triangle $= \frac{1}{2}bh = \frac{1}{2} \times 3.26 \times 7.5 = 12.225 \text{ cm}^2$ (by long multiplication or by calculator).

The approximate value is $\frac{1}{2} \times 3 \times 8 = 12$ cm^2, so there are no obvious blunder or order of magnitude errors. However it is not usual in a measurements type problem to state the answer to an accuracy greater than 1 significant figure more than the least accurate number in the data; this is 7.5 cm, so the result should not have more than 3 significant figures.
Thus area of triangle = 12.2 cm^2.

Problem 2 State which type of error has been made in the following statements: (a) $72 \times 31.429 = 2262.9$, (b) $16 \times 0.08 \times 7 = 89.6$, (c) $11.714 \times 0.0088 = 0.3247$ correct to 4 decimal places, (d) $\frac{29.74 \times 0.0512}{11.89} = 0.12$, correct to 2 significant figures.

(a) $72 \times 31.429 = 2262.888$ (by long multiplication or calculator); hence a **rounding-off error** has occurred. The answer should have stated: $72 \times 31.429 = 2262.9$ correct to 5 significant figures.

(b) $16 \times 0.08 \times 7 = {}^4\cancel{16} \times \frac{8}{\cancel{100}_{25}} \times 7 = \frac{32 \times 7}{25} = \frac{224}{25} = 8\frac{24}{25} = 8.96.$

Hence an **order of magnitude error** has occurred.

(c) 11.714×0.0088 is approximately equal to $12 \times 9 \times 10^{-3}$, i.e., about 108×10^{-3} or 0.108. Thus a **blunder** has been made.

(d) $\frac{29.74 \times 0.0512}{11.89} \simeq \frac{30 \times 5 \times 10^{-2}}{12} \simeq \frac{150}{12 \times 10^2} \simeq \frac{15}{120} \simeq \frac{1}{8}$ or 0.125, hence, no order of magnitude error has occurred. However, $\frac{29.74 \times 0.0512}{11.89} = 0.128$, correct to 3 significant figures, which equals 0.13 correct to 2 significant figures. Hence a **rounding-off error** has occurred.

Problem 3 Find from 4-figure tables the values of (a) lg 5.279, (b) lg 378.2, (c) lg 0.004 831.

(a) $\lg 5.279 = \lg (5.279 \times 10^0)$
The characteristic (i.e. integer) is given by the power of 10, in this case, 0.
The mantissa (i.e. decimal part) is read from logarithm tables.
Hence $\lg (5.279 \times 10^0) =$ **0.7225**

(b) $\lg 378.2 = \lg (3.782 \times 10^2) =$ **2.5777**

(c) $\lg 0.004\,831 = \lg (4.831 \times 10^{-3}) = -3 + 0.6840 =$ $\mathbf{\bar{3}.6840}$

Problem 4 Find from 4-figure tables the antilogarithms of (a) 0.2783, (b) 3.8329, (c) $\bar{1}.6377$.

(a) Only the decimal part is used when reading antilogarithm tables. 0.2783 from antilogarithm tables corresponds to 1.898 (i.e. a number between 1.0 and 9.999).
The characteristic 0 indicates that 1.898 is multiplied by 10^0, (i.e. 1). Hence the antilogarithm of 0.2783 is **1.898**

(b) From antilogarithm tables, 0.8329 corresponds to 6.806. The characteristic 3 indicates that 6.806 is multiplied by 10^3. Hence the antilogarithm of 3.8329 is 6.806×10^3, i.e. **6806**

(c) The antilogarithm of $\bar{1}.6377$ is 4.342×10^{-1}, i.e., **0.4342**

Problem 5 Use logarithms to find the value of 26.47×0.02981

For multiplication, the logarithms are added, i.e. $\lg (A \times B) = \lg A + \lg B$.

Number	*Logarithm*	
$26.47 = 2.647 \times 10^1$ $0.02981 = 2.981 \times 10^{-2}$	1.4227 $\bar{2}.4743$	
7.889×10^{-1}	$\bar{1}.8970$	Adding

Hence $26.47 \times 0.02981 = 7.889 \times 10^{-1} =$ **0.7889**

Problem 6 Use logarithms to evaluate $46.32 \times 97.17 \times 0.1258$

For multiplication, the logarithms are added, no matter how many numbers involved.

Number	*Logarithm*	
$46.32 = 4.632 \times 10^1$ $97.17 = 9.717 \times 10^1$ $0.1258 = 1.258 \times 10^{-1}$	1.6658 1.9875 $\bar{1}.0996$	
5.661×10^2	2.7529	Adding

Hence $46.32 \times 97.17 \times 0.1258 = 5.661 \times 10^2 =$ **566.1**

Problem 7 Use logarithms to evaluate $\dfrac{4.621}{23.76}$

When numbers are divided, the logarithms are subtracted, i.e. $\lg \left(\dfrac{A}{B}\right) = \lg A - \lg B$.

Number	*Logarithm*	
$4.621 = 4.621 \times 10^0$ $23.76 = 2.376 \times 10^1$	0.6647 1.3758	
1.945×10^{-1}	$\bar{1}.2889$	Subtracting

Hence $\dfrac{4.621}{23.76} = 1.945 \times 10^{-1} =$ **0.1945**

Problem 8 Use logarithms to evaluate $\dfrac{1}{52.73}$

Number	Logarithm	
	$\bar{1}$ 9 9 10	
$1 = 1 \times 10^0$	$\not{0}.\not{0}\not{0}\not{0}\ 0$	
$52.73 = 5.273 \times 10^1$	1.722 0	
1.897×10^{-2}	$\bar{2}$.278 0	Subtracting

(Note, if 1 is 'borrowed' from 0 then $\bar{1}$ is left. Alternatively, add 1 to top and bottom).

Hence $\frac{1}{52.73} = 1.897 \times 10^{-2} =$ **0.018 97** (which may be checked using reciprocal tables).

Problem 9 Use logarithms to evaluate $(31.28)^3$.

To find the n'th power of a number, the logarithm of a number is multiplied by n, i.e., $\lg A^n = n \lg A$.

Number	Logarithm	
$31.28 = 3.128 \times 10^1$	1.4953 3	
3.061×10^4	4.4859	Multiplying

Hence $(31.28)^3 = 3.061 \times 10^4 =$ **30610**

Problem 10 Evaluate, using logarithms $(0.8316)^4$.

Number	Logarithm	
$0.8316 = 8.316 \times 10^{-1}$	$\bar{1}$.9199 4	
$= 4.782 \times 10^{-1}$	$\bar{1}$.6796	Multiplying

Hence $(0.8316)^4 = 4.782 \times 10^{-1} = 0.4782$

Problem 11 Determine $\sqrt{546.2}$ using logarithms.

Since $\sqrt{546.2} = (546.2)^{1/2}$ then the logarithm of 546.2 is multiplied by $\frac{1}{2}$, which is the same as dividing by 2.

Number	Logarithm
$546.2 = 5.462 \times 10^2$ 2.337×10^1	2) 2.7374 1.3687

Hence $\sqrt{546.2} = \pm 2.337 \times 10^1 =$ **±23.37**

Problem 12 Find the value of $\sqrt{0.007\ 328}$ using logarithms.

Number	*Logarithm*
$0.007\ 328 = 7.328 \times 10^{-3}$	$\bar{3}.8650$
	$2\,)\ \bar{4} + 1.8650$
8.561×10^{-2}	$\bar{2} + 0.9325$

If the characteristic $\bar{3}$ is divided by 2 it does not give an exact integer answer. Hence -1 is added to $\bar{3}$ to make $\bar{4}$ and 1 is added to the mantissa to make the overall value of the logarithm the same, i.e. $\bar{3}.8650$ is the same as $\bar{4} + 1.8650$

Hence $\sqrt{0.007\ 328} = 8.561 \times 10^{-2}$ = **0.08561**

Problem 13 Evaluate $\left(\dfrac{19.74 \times 0.0861}{3.462}\right)^4$ using logarithms.

Number	*Logarithm*	
$19.74 = 1.974 \times 10^1$	1.2953	
$0.0861 = 8.61 \times 10^{-2}$	$\bar{2}.9350$	
	$\bar{1}\ 1\ 2\ 9\ 1$	
	$\not{0}.\not{2}\,\not{3}\,\not{0}\,3$	Adding
$3.462 = 3.462 \times 10^0$	0.5394	
	$\bar{1}.6909$	Subtracting
	4	
5.802×10^{-2}	$\bar{2}.7636$	Multiplying

Hence $\left(\dfrac{19.74 \times 0.0861}{3.462}\right)^4 = 5.802 \times 10^{-2}$ = **0.05802**

Problem 14 Use logarithms to find the energy E where $E = \frac{1}{2} L I^2$ joules given that current $I = 4.78$ amperes and inductance $L = 0.342$ henry.

$$E = \frac{1}{2} L I^2 = \frac{1}{2} \times 0.342 \times 4.78^2$$

Number	*Logarithm*	
$I = 4.78 = 4.78 \times 10^0$	0.6794	
	2	
I^2	1.3588	Multiplying
$L = 0.342 = 3.42 \times 10^{-1}$	$\bar{1}.5340$	
$LI^2 = 7.812 \times 10^0$	0.8928	Adding

Hence $E = \frac{1}{2} L I^2 = \frac{1}{2} \times 7.812 \times 10^0 = \frac{7.812}{2}$ = **3.906 joules**

Problem 15 Kinetic energy E is given by $E = \frac{1}{2}m(v^2 - u^2)$. Use logarithms to evaluate E when mass $m = 220$ kg, final velocity $v = 72.3$ m/s and the initial velocity $u = 33.7$ m/s.

Addition and subtraction **cannot** be achieved using logarithms. v^2 and u^2 may be determined using logarithms, but the value of $(v^2 - u^2)$ must be determined by subtraction.

Number	*Logarithm*
$v = 72.3 = 7.23 \times 10^1$	1.8591 2
$v^2 = 5226$	3.7182

Number	*Logarithm*
$u = 33.7 = 3.37 \times 10^1$	1.5276 2
$u^2 = 1136$	3.0552

Thus $v^2 - u^2 = 5226 - 1136 = 4090$

Number	*Logarithm*
$v^2 - u^2 = 4090 = 4.09 \times 10^3$ $m = 220 = 2.20 \times 10^2$	3.6117 2.3424
$m(v^2 - u^2) = 8.997 \times 10^5$	5.9541

Thus $\frac{1}{2}m(v^2 - u^2) = \frac{8.997 \times 10^5}{2} = \frac{899\,700}{2} = 449\,850$ joules.

From the data given the answer should be expressed to no more than 4 significant figures. **Thus kinetic energy E = 449 900 joules.**

Problem 16 Impedance Z is given by $Z = \sqrt{(R^2 + X_c^2)}$ where $X_c = \frac{1}{2\pi fC}$.

Evaluate Z when $R = 40$, $C = 60 \times 10^{-6}$, $f = 50$ and $\pi = 3.142$.

Number	*Logarithm*
2 3.142 50 60×10^{-6}	0.3010 0.4972 1.6990 $\bar{5}.7782$
$2\pi fC$	$\bar{2}.2754$

Number	*Logarithm*
1 $2\pi fC$	0.0000 $\bar{2}.2754$
$X_c = \frac{1}{2\pi fC}$	1.7246 2
$X_c^2 = 2813$	3.4492

$R^2 + X_c^2 = 40^2 + 2813 = 1600 + 2813 = 4413.$

Number	Logarithm
$R^2 + X_c^2$	2) 3.6447
$\sqrt{(R^2 + X_c^2)}$	1.8224

Hence $\sqrt{(R^2 + X_c^2)} = \mathbf{66.43} = Z$

C. FURTHER PROBLEMS INVOLVING ARITHMETIC CALCULATIONS

1 The area A of a rectangle is given by $A = lb$. The length l when measured is found to be 23.1 mm and the breadth b is 7.8 mm. Determine the area of the rectangle.
[180 mm²]

2 The velocity of a body is given by $v = u + at$. The initial velocity u is measured when time t is 15 seconds and found to be 12 m/s. If the acceleration a is 9.81 m/s², calculate the final velocity v.
[159 m/s]

3 Calculate the current I in an electrical circuit, where $I = \frac{V}{R}$ amperes when the voltage V is measured and found to be 7.2 volts and the resistance R is 17.7 ohms.
[0.407 amperes]

In *Problems 4 to 8* state which type of error, or errors, have been made.

4 $25 \times 0.06 \times 1.4 = 0.21$. [order of magnitude error]

5 $137 \times 6.842 = 937.4$ [rounding-off error – should add 'correct to 4 significant figures'.]

6 $\frac{204 \times 0.008}{12.6} = 10.42$ [Blunder]

7 For a gas $pV = c$. When pressure $p = 103\,400$ Pa and $V = 0.54$ m³ then $c = 55\,836$ Pa m³.
[Measured values, hence $c = 55\,800$ Pa m³]

8 $\frac{4.6 \times 0.07}{52.3 \times 0.274} = 0.225$
[Order of magnitude error and rounding-off error – should be 0.0225 correct to 3 significant figures]

In *Problems 9 and 10*, use 4-figure tables to evaluate the quantities shown.

9 (a) lg 3.764; (b) lg 241.8, (c) lg 1.0; (d) lg 0.07632.
[(a) 0.5757; (b) 2.3834; (c) 0; (d) $\bar{2}.8826$]

10 (a) antilogarithm of 0.5862; (b) antilogarithm of 4.731; (c) antilogarithm of $\bar{2}.3197$.
[(a) 3.857; (b) 53830; (c) 0.02087]

In *Problems 11 to 29*, use 4-figure logarithms to evaluate.

11 (a) 43.27×12.91; (b) 54.31×0.5724.
[(a) 558.6. (b) 31.09]

12 (a) $127.8 \times 0.0431 \times 19.8$; (b) $15.76 \div 4.329$.
[(a) 109.1; (b) 3.641]

13 (a) $\frac{137.6}{552.9}$; (b) $\frac{11.82 \times 1.736}{0.041}$.
[(a) 0.2489; (b) 500.3]

14 (a) $\frac{1}{17.31}$, (b) $\frac{1}{0.0346}$; (c) $\frac{1}{147.9}$.
[(a) 0.05777; (b) 28.90; (c) 0.006763]

15 (a) 13.6^3; (b) 3.476^4; (c) 0.124^5.

[(a) 2515; (b) 146.0; (c) 0.00002931]

16 (a) $\sqrt{347.1}$; (b) $\sqrt{7632}$; (c) $\sqrt{0.027}$; (d) $\sqrt{0.004168}$

[(a) 18.63; (b) 87.36; (c) 0.1644; (d) 0.06455]

17 (a) $\left(\dfrac{24.68 \times 0.0532}{7.412}\right)^3$; (b) $\left(\dfrac{0.2681 \times 41.2^2}{32.6 \times 11.89}\right)^4$.

[(a) 0.005558; (b) 1.900]

18 (a) $\dfrac{14.32^3}{21.68^2}$, (b) $\dfrac{4.821^3}{17.33^2 - 15.86 \times 11.6}$.

[(a) 6.244; (b) 0.9668]

19 Find the distance s, given that $s = \frac{1}{2} g\, t^2$. Time $t = 0.032$ seconds and acceleration due to gravity $g = 9.81$ m/s^2.

[0.00502 m or 5.02 mm]

20 The energy stored in a capacitor is given by $E = \frac{1}{2} C\, V^2$ joules. Determine the energy when capacitance $C = 5 \times 10^{-6}$ farads and voltage $V = 240$ V.

[0.1440 joules]

21 Resistance R_2 is given by $R_2 = R_1\,(1 + \alpha t)$. Find R_2, correct to 4 significant figures when $R_1 = 220$, $\alpha = 0.00027$ and $t = 75.6$.

[224.5]

22 Density $= \dfrac{\text{mass}}{\text{volume}}$. Find the density when the mass is 2.462 kg and the volume is 173 cm^3. Give the answer in units of kg/m^3.

[14230 kg/m^3]

23 Velocity = frequency × wavelength. Find the velocity when the frequency is 1825 Hz and the wavelength is 0.154 m.

[281.0 m/s]

24 Evaluate resistance R_T, given $\dfrac{1}{R_T} = \dfrac{1}{R_1} + \dfrac{1}{R_2} + \dfrac{1}{R_3}$, when $R_1 = 5.5\Omega$, $R_2 = 7.42\Omega$ and $R_3 = 12.6\Omega$.

[2.526Ω]

25 Find the total cost of 37 calculators costing £12.65 and 19 slide rules costing £6.38 each.

[£589.30]

26 Power $= \dfrac{\text{force} \times \text{distance}}{\text{time}}$. Find the power when a force of 3760 N raises an object a distance of 4.73 m in 35 seconds.

[508.2 watts]

27 In a geometric progression, $a\,r^5 = 73.21$. Evaluate r when a is 3.478.

[1.839]

28 Impedance Z is given by $Z = \sqrt{(R^2 + X_L{}^2)}$ where $X_L = 2\pi f L$. Evaluate Z when $R = 32.4$, $f = 100$, $L = 40 \times 10^{-3}$ and $\pi = 3.142$.

[41.01]

29 Below is a table of some metric to imperial conversions.

Length:	2.54 cm	= 1 inch
	1.61 km	= 1 mile
Weight:	1 kg	= 2.2 lb (1 lb = 16 ounces)
Capacity:	1 litre	= 1.76 pints (8 pints = 1 gallon)

Use the table to determine (a) the number of millimetres in 15 inches, (b) a speed of 35 m.p.h. in km/h, (b) the number of kilometres in 235 miles, (d) the number of pounds and ounces in 24 kg (correct to the nearest ounce), (e) the number of kilograms in 15 lbs, (f) the number of litres in 12 gallons, and (g) the number of gallons in 25 litres.

[(a) 381 mm; (b) 56.35 km/h; (c) 376 km; (d) 52 lbs 13 ozs; (e) 6.82 kg; (f) 54.55 l; (g) 5.5 gallons.]

30 In *Problems 11 to 29*, evaluate using a calculator and/or slide rule.

8 Exponential functions and Naperian logarithms

A. MAIN POINTS CONCERNED WITH EXPONENTIAL FUNCTIONS AND NAPERIAN LOGARITHMS

1 An **exponential function** is one which contains e^x, e being a constant called the exponent and having an approximate value of 2.7183. The exponent arises from the natural laws of growth and decay and is used as a base of natural or Naperian logarithms.

2 **The natural laws of growth and decay** are of the form $y = A\ e^{kx}$, where A and K are constants. The natural laws occur frequently in engineering and science and examples of quantities related by a natural law include:

(i) Linear expansion $l = l_0\ e^{\alpha\theta}$
(ii) Change in electrical resistance with temperature $R_\theta = R_0\ e^{\alpha\theta}$
(iii) Tension in belts $T_1 = T_0\ e^{\mu\alpha}$
(iv) Newton's law of cooling $\theta = \theta_0\ e^{-kt}$
(v) Biological growth $y = y_0\ e^{kt}$
(vi) Discharge of a capacitor $q = Qe^{-\frac{t}{CR}}$
(vii) Atmospheric pressure $p = p_0\ e^{-\frac{h}{c}}$
(vii) Radioactive decay $N = N_0\ e^{-\lambda t}$
(iv) Decay of current in an inductive circuit $i = I\ e^{-\frac{Rt}{L}}$

3 The **value** of e^x may be determined by using

(i) a calculator which possesses an 'e^x' function,
(ii) the power series $e^x = 1 + x + \dfrac{x^2}{2!} + \dfrac{x^3}{3!} + \ldots\ldots$ (where 3! is 'factorial 3' and means $3 \times 2 \times 1$),
(iii) Naperian logarithms (see paras. 4 to 6), or
(iv) 4 figure tables of exponential functions which enable values of e^x and e^{-x} to be read over a range of x from 0.02 to 6.0. For example, $e^{0\cdot36} = 1.4333$, $e^{4\cdot3} = 73.700$, $e^{-0\cdot47} = 0.6250$ and $e^{-2\cdot6} = 0.0743$. Also, from tables and laws of indices, $e^{0\cdot68} = e^{0\cdot6 + 0\cdot08} = (e^{0\cdot6})(e^{0\cdot08}) = (1.8221)(1.0833)$
$= 1.9739$ correct to 4 decimal places.

(See *Problems 1 and 2*)

Hyperbolic or Naperian logarithms

4 (i) **A logarithm** of a number is the power to which a base has to be raised to be equal to the number.
Thus, if $y = a^x$ then $x = \log_a y$.

(ii) Logarithms having a base of 10 are called **common logarithms** and the common logarithm of x is written as $\lg x$. (see chapter 7)

(iii) Logarithms having a base of e are called **hyperbolic, Naperian** or **natural logarithms** and the Naperian logarithm of x is written as $\log_e x$, or more commonly, $\ln x$.

5 The **value** of a **Naperian logarithm** may be obtained by:

(i) a calculator possessing a '$\ln x$' function,

(ii) the change of base rule for logarithms, which states, $\log_a y = \dfrac{\log_b y}{\log_b a}$, from which, $\ln y = \dfrac{\lg y}{\lg e} = \dfrac{\lg y}{0.4343} = 2.3026 \lg y$, or

(iii) 4 figure Naperian logarithm tables.

6 **Use of 4 figure Naperian logarithm tables**

(i) **For numbers from 1 to 10,** Naperian logarithm tables are used in a similar way to common logarithms. For example,

$\ln 3.4 = 1.2238$

$\ln 3.47 = 1.2442$

$\lg 3.478 = 1.2442 + 23$ (from the mean difference column equivalent to 8)

i.e. $\ln 3.478 = 1.2465$

Similarly, $\ln 5.731 = 1.7459$ and $\ln 9.159 = 2.2148$

(ii) **Numbers larger than 10** are initially expressed in standard form and the supplementary table of Naperian logarithms of 10^{+n} used.

For example, $\ln 64 = \ln(6.4 \times 10^1) = \ln 6.4 + \ln 10^1$
$= 1.8563 + 2.3026 = 4.1589$

Similarly, $\ln 327.6 = \ln(3.276 \times 10^2) = \ln 3.276 + \ln 10^2$
$= 1.1866 + 4.6052 = 5.7918$

(iii) **Numbers smaller than 1** are initially expressed in standard form and the supplementary table of Naperian logarithms of 10^{-n} used.

For example, $\ln 0.064 = \ln(6.4 \times 10^{-2}) = \ln 6.4 + \ln 10^{-2}$
$= 1.8563 + \bar{5}.3948 = \bar{3}.2511$

$\bar{3}.2511$ is the same as $-3 + 0.2511 = -2.7489$ (which is the value given by a calculator)

Similarly, $\ln 0.003276 = \ln(3.276 \times 10^{-3}) = \ln 3.276 + \ln 10^{-3}$
$= 1.1866 + \bar{7}.0922 = \bar{6}.2788$ or -5.7212

The value of a Naperian logarithm of a positive number less than unity is always wholly negative.

(iv) **The antilogarithm of a number between 0 and 2.3026** is obtained by finding the value of the logarithm within the table, the anti-logarithm being obtained from the corresponding row and column. For example, if $\ln x = 1.4317$, the nearest number less than this is 1.4303 corresponding to 4.18. The difference between 1.4317 and 1.4303 is 14, which in the mean difference column corresponds to 6.

Hence, if $\ln x = 1.4317$, then $x = 4.186$

(v) **To find the antilogarithm of a number greater than 2.3026**, for example, 5.3726, the nearest number less than 5.3726 is found in the supplementary table of Naperian logarithms of 10^{+n}; in this case it is 4.6052, corresponding to 10^2. The

difference between 5.3726 and 4.6052 is 0.7674. The antilogarithm of 0.7674 is read directly from within the table as 2.154.
Hence, if $\ln x = 5.3726$ then $x = 2.154 \times 10^2 = 215.4$

(vi) **To find the antilogarithm of a number less than 2.3026**, for example, -5.8314, the number is initially change to $\bar{6}.1686$ and the nearest more negative number in the supplementary table of Naperian logarithms of 10^{-n} is found; in this case it is $\bar{7}.0922$, corresponding to 10^{-3}. The difference between $\bar{6}.1686$ and $\bar{7}.0922$ is 1.0764. The antilogarithm of 1.0764 is read directly from within the tables as 2.934.
Hence, if $\ln x = -5.8314$ then $x = 2.934 \times 10^{-3} = 0.002934$

(vii) The procedure for determining antilogarithms is considerably quicker with a **calculator.** If $\ln x = y$ then $x = e^y$, from the definition of a logarithm.
Thus, if $\ln x = 5.3726$, then $x = e^{5.3726} = 215.42224$, by calculator and if $\ln x = -5.8314$ then $x = e^{-5.8314} = 2.9339 \times 10^{-3} = 0.0029339$, by calculator.
Alternatively, an 'invert ln *x*' function is available on many calculators.
(See *Problems 3 to 6*)

7 It is sometimes necessary to transpose an equation of the form $V = A\, e^{bx}$ to find x. This is achieved as follows:

Rearrange the equation to leave the exponent on one side on its own,

i.e. $e^{bx} = \dfrac{V}{A}$

Taking Naperian logarithms of both sides gives:

$$\ln (e^{bx}) = \ln (\frac{V}{A})$$

i.e. $b\,x \ln e = \ln (\frac{V}{A})$, from the laws of logarithms

i.e. $b\,x = \ln (\frac{V}{A})$, since $\ln e = 1$

Hence $x = \dfrac{1}{b} \ln (\dfrac{V}{A})$.

(See *Problems 7 to 10*)

B. WORKED PROBLEMS ON EXPONENTIAL FUNCTIONS AND NAPERIAN LOGARITHMS

Problem 1 Use exponential tables to determine the values of (a) $e^{0.18}$, (b) $e^{-1.4}$, (c) $3\,e^{-0.54}$, (d) $0.2\,e^8$, each correct to 4 significant figures.

From exponential tables:
(a) $e^{0.18} = 1.1972 = \mathbf{1.197}$, correct to 4 significant figures.
(b) $e^{-1.4} = \mathbf{0.2466}$
(c) $e^{-0.54} = (e^{-0.5})(e^{-0.04}) = (0.6065)(0.9608) = 0.58273$
Hence $3e^{-0.54} = 3\,(0.58273) = \mathbf{1.748}$, correct to 4 significant figures.
(d) $e^8 = e^{4+4} = (e^4)(e^4) = (54.598)^2 = 2980.94$
Hence $0.2\,e^8 = (0.2)(2980.94) = \mathbf{596.2}$, correct to 4 significant figures.

Problem 2 Determine the value of $5\,e^{0.5}$, correct to 5 significant figures by using the power series for e^x.

$$e^x = 1 + x + \frac{x^2}{2!} + \frac{x^3}{3!} + \ldots\ldots$$

Hence $e^{0.5} = 1 + 0.5 + \frac{(0.5)^2}{(2)(1)} + \frac{(0.5)^3}{(3)(2)(1)} + \frac{(0.5)^4}{(4)(3)(2)(1)} + \frac{(0.5)^5}{(5)(4)(3)(2)(1)} + \frac{(0.5)^6}{(6)(5)(4)(3)(2)(1)}$

$= 1 + 0.5 + 0.125 + 0.020833 + 0.0026042 + 0.0002604 + 0.0000217$

i.e. $e^{0.5} = 1.64872$, correct to 6 significant figures.

Hence $5e^{0.5} = 5\,(1.64872) = $ **8.2436**, correct to 5 significant figures.

Problem 3 Use tables to determine (a) ln 7.483; (b) ln 748.3; (c) ln 0.007483.

(a) ln 7.4 = 2.0015
ln 7.48 = 2.0122
ln 7.483 = 2.0122 + 4 (from the mean difference column equivalent to 3)
i.e. ln 7.483 = **2.0126**

(b) From para. 6(ii), $\ln 748.3 = \ln (7.483 \times 10^2) = \ln 7.483 + \ln 10^2$
$= 2.0126 + 4.6052 =$ **6.6178**

(c) From para. 6(iii), $\ln 0.007483 = \ln (7.483 \times 10^{-3}) = \ln 7.483 + \ln 10^{-3}$
$= 2.0126 + \overline{7}.0922 = \mathbf{\overline{5}.1048}$ **or −4.8952**

Problem 4 Determine from tables the value of x when $\ln x$ is (a) 1.7321; (b) 8.4628, (c) −8.4217.

(a) Since 1.7321 lies between 0 and 2.3026 the value of x can be read directly from within the tables of Naperian logarithms. The nearest number less than 1.7321 is 1.7317 corresponding to 5.65. The difference between 1.7321 and 1.7317 is 4, which corresponds to 2 in the mean difference column.
Hence if $\ln x = 1.7321$, then $x =$ **5.652**
(Alternatively, if $\ln x = 1.7321$, then $x = e^{1.7321} = 5.6525117$ by calculator – or an 'invert ln x' is available on many calculators.)

(b) The nearest number less than 8.4628 in the supplementary table of Naperian logarithms of 10^{+n} is 6.9078 corresponding to 10^3. The difference between 8.4628 and 6.9078 is 1.5550. The antilogarithm of 1.5550 is found from within the tables and is 4.735.
Hence, if $\ln x = 8.4628$, then $x = 4.735 \times 10^3 =$ **4735.**
(Alternatively, $x = e^{8.4628} = 4735.2983$ by calculator.)

(c) $-8.4217 = -9 + 0.5783 = \overline{9}.5783$.
The nearest more negative number to $\overline{9}.5783$ in the supplementary table of Naperian logarithms of 10^{-n} is $\overline{10}.7897$, corresponding to 10^{-4}. The difference between 9.5783 and $\overline{10}.7897$ is 0.7886. The antilogarithm of 0.7886 is found from within the tables and is 2.200.
Hence if $\ln x = -8.4217$, then $x = \mathbf{2.200 \times 10^{-4}}$.

Problem 5 Use common logarithms to evaluate ln 36.25.

From the change of base rule, $\ln y = 2.3026 \lg y$ (from para. 5).
Hence ln 36.25 = (2.3026)(lg 36.25)
= (2.3026)(1.5593) = **3.5904**, correct to 4 decimal places,
which may be checked by using Naperian logarithms or a calculator.

Problem 6 Use Naperian logarithms to evaluate (a) $e^{4.137}$ and (b) $-4.5\,e^{-2.4379}$, each correct to 3 significant sigures.

(a) Let $y = e^{4.137}$. Taking Naperian logarithms of each side gives:

$\ln y = \ln e^{4.137} = 4.137 \ln e$ by the laws of logarithms.

Thus $\ln y = 4.137$, since $\ln e = 1$.

$$\ln y = 2.3026 + (4.137 - 2.3026)$$
$$= 2.3026 + 1.8344.$$

Taking antilogarithms of both sides gives $y = 10^1 \times 6.261$

i.e. $e^{4.137} = \mathbf{62.6}$, correct to 3 significant figures.

(b) Let $y = e^{-2.4379}$. Taking Naperian logarithms of each side gives:

$$\ln y = -2.4379 = \bar{3}.5621 = \bar{5}.3948 + (\bar{3}.5621 - \bar{5}.3948)$$
$$\text{i.e. } \ln y = \bar{5}.3948 + 2.1673$$

Taking antilogarithms of both sides gives $y = 10^{-2} \times 8.734$.

Thus $-4.5\,e^{-2.4379} = (-4.5)(8.734 \times 10^{-2}) = \mathbf{-0.393}$, correct to 3 significant figures.

Problem 7 The tension in a belt is given by the equation $T_1 = T_0\,e^{\mu\alpha}$. Determine α when $T_1 = 30.0$, $T_0 = 24.0$ and $\mu = 0.25$.

Transposing $T_1 = T_0\,e^{\mu\alpha}$ gives $e^{\mu\alpha} = \dfrac{T_1}{T_0}$

Taking Naperian logarithms of both sides gives:

$$\ln e^{\mu\alpha} = \ln \frac{T_1}{T_0}$$

$$\text{i.e. } \mu\alpha \ln e = \ln \frac{T_1}{T_0}, \text{ from the laws of logarithms}$$

$$\text{i.e. } \mu\alpha = \ln \frac{T_1}{T_0}, \text{ since } \ln e = 1$$

$$\text{Hence } \alpha = \frac{1}{\mu} \ln\left(\frac{T_1}{T_0}\right).$$

$T_1 = 30.0$, $T_0 = 24.0$ and $\mu = 0.25$,

$$\text{thus } \alpha = \frac{1}{0.25} \ln\left(\frac{30.0}{24.0}\right) = \mathbf{0.8926}$$

Problem 8 The resistance R of an electrical conductor at temperature θ°C is given by $R = R_0\,e^{\alpha\theta}$, where α is a constant and $R_0 = 5 \times 10^3$ ohms. Determine the value of α when $R = 6 \times 10^3$ ohms and $\theta = 1500$°C. Also, find the temperature when the resistance R is 5.4×10^3 ohms.

Transposing $R = R_0 e^{\alpha\theta}$ gives $\dfrac{R}{R_0} = e^{\alpha\theta}$

Taking Naperian logarithms of both sides gives: $\ln\left(\dfrac{R}{R_0}\right) = \ln e^{\alpha\theta} = \alpha\theta$

$$\text{Hence } \alpha = \frac{1}{\theta} \ln\left(\frac{R}{R_0}\right) = \frac{1}{1500} \ln\left(\frac{6 \times 10^3}{5 \times 10^3}\right) = \frac{1}{1500}(0.1823)$$

Hence $\alpha = \mathbf{1.215 \times 10^{-4}}$ (by calculator or tables).

From above, $\ln\left(\dfrac{R}{R_0}\right) = \alpha\theta$, hence $\theta = \dfrac{1}{\alpha} \ln\left(\dfrac{R}{R_0}\right)$

When $R = 5.4 \times 10^3$, $\alpha = 1.215 \times 10^{-4}$ and $R_0 = 5 \times 10^3$

$$= \frac{1}{1.215 \times 10^{-4}} \ln\left(\frac{5.4 \times 10^3}{5 \times 10^3}\right)$$

$$= \frac{10^4}{1.215}(7.696 \times 10^{-2}) = \mathbf{633.4°C}.$$

Problem 9 In an experiment involving Newton's law of cooling, the temperature θ (°C) is given by $\theta = \theta_0\, e^{-kt}$. Find the value of constant k when $\theta_0 = 56.6°C$, $\theta = 16.5°C$ and $t = 83.0$ seconds.

Transposing $\theta = \theta_0 e^{-kt}$ gives $\frac{\theta}{\theta_0} = e^{-kt}$ from which $\frac{\theta_0}{\theta} = \frac{1}{e^{-kt}} = e^{kt}$

Taking Naperian logarithms of both sides gives: $\ln\left(\frac{\theta_0}{\theta}\right) = kt$,

from which, $k = \frac{1}{t}\ln\left(\frac{\theta_0}{\theta}\right) = \frac{1}{83.0}\ln\left(\frac{56.6}{16.5}\right) = \frac{1}{83.0}(1.2326)$

Hence $k = \mathbf{1.485 \times 10^{-2}}$.

Problem 10 The current i amperes flowing in a capacitor at time t seconds is given by $i = 8.0(1 - e^{-\frac{t}{CR}})$, where the circuit resistance R is 25×10^3 ohms and capacitance C is 16×10^{-6} farads. Determine (a) the current i after 0.5 seconds and (b) the time for the current to reach 6.0 amperes.

(a) Current $i = 8.0(1 - e^{-\frac{t}{CR}}) = 8.0\left[1 - e^{\left(-\frac{0.5}{(16 \times 10^{-6})(25 \times 10^3)}\right)}\right] = 8.0(1 - e^{-1.25})$

$= 8.0(1 - 0.2865) = 8.0(0.7135) = \mathbf{5.708\ amperes}.$

(b) Transposing $i = 8.0(1 - e^{-\frac{t}{CR}})$ gives $\frac{i}{8.0} = 1 - e^{-\frac{t}{CR}}$,

from which, $e^{-\frac{t}{CR}} = 1 - \frac{i}{8.0} = \frac{8.0 - i}{8.0}$

Taking the reciprocal of both sides gives: $e^{\frac{t}{CR}} = \frac{8.0}{8.0 - i}$

Taking Naperian logarithms of both sides gives: $\frac{t}{CR} = \ln\left(\frac{8.0}{8.0 - i}\right)$

Hence $t = CR\ln\left(\frac{8.0}{8.0 - i}\right) = (16 \times 10^{-6})(25 \times 10^3)\ln\left(\frac{8.0}{8.0 - 6.0}\right)$

when $i = 6.0$ amperes,

i.e. $t = \frac{400}{10^3}\ln\left(\frac{8.0}{2.0}\right) = 0.4 \ln 4.0 = 0.4(1.3863) = \mathbf{0.5545\ seconds}.$

C. FURTHER PROBLEMS ON EXPONENTIAL FUNCTIONS AND NAPERIAN LOGARITHMS

In *Problems 1 and 2* use exponential tables to evaluate the given functions correct to 4 significant figures. Check your results by using a calculator.

1 (a) $e^{4.4}$; (b) $e^{-0.25}$; (c) $e^{0.92}$.

[(a) 81.45; (b) 0.7788; (c) 2.509]

2 (a) $e^{-1.8}$; (b) $e^{-0.78}$; (c) e^{10}.

[(a) 0.1653; (b) 0.4584; (c) 22030]

3 Use the power series for e^x to determine, correct to 4 significant figures, (a) e^2, (b) $e^{-0.3}$ and check your result by using exponential tables and/or a calculator.

[(a) 7.389; (b) 0.7408]

4 Evaluate, correct to 4 significant figures, (a) $3.5e^{2.8}$, (b) $-\frac{6}{5}e^{-1.5}$ (c) $2.16e^{5.7}$.

[(a) 57.56; (b) –0.2678; (c) 645.6]

In *Problems 5 to 7* use 4 figure tables to evaluate the given functions.

5 (a) ln 1.73; (b) ln 5.413; (c) ln 9.412.

[(a) 0.5481; (b) 1.6887; (c) 2.2420]

6 (a) ln 17.3; (b) ln 541.3; (c) ln 9412.

[(a) 2.8507; (b) 6.2939; (c) 9.1498]

7 (a) ln 0.173; (b) ln 0.005413, (c) ln 0.09412.

[(a) $\bar{2}.2455$ or –1.7545; (b) $\bar{6}.7809$ or –5.2191; (c) $\bar{3}.6368$ or –2.3632]

In *Problems 8 to 10* use 4 figure tables to find the value of x when ln x has the value shown.

8 (a) 0.4317; (b) 1.2047; (c) 2.0491.

[(a) 1.540; (b) 3.336; (c) 7.761]

9 (a) 3.1429; (b) 6.3312; (c) 10.3171.

[(a) 23.17; (b) 561.8; (c) 30250]

10 (a) $\bar{1}.3617$; (b) $\bar{6}.3173$; (c) –3.3117.

[(a) 0.5282; (b) 0.003404; (c) 0.03645]

11 Use common logarithms to evaluate (a) ln 29.41 and (b) ln 110.5, each correct to 4 significant figures.

[(a) 3.381; (b) 4.705]

12 Use Naperian logarithms to evaluate, correct to 3 significant figures, (a) $e^{3.912}$; (b) $2.7e^{-1.8463}$; (c) $-\frac{15}{8}e^{2.3131}$.

[(a) 50.0; (b) 0.426; (c) –18.9]

13 Evaluate $0.26e^{-\frac{3x}{7}}$ when x has a value of (a) –3.68; (b) 2.417; (c) 14.

[(a) 1.2587; (b) 0.09228; (c) 6.445×10^{-4}]

14 Transpose (a) $y = y_0e^{kt}$ for t; (b) $N = N_0e^{-\lambda t}$ for λ.

$$\left[\begin{array}{l}\text{(a) } t = \frac{1}{k}\ln\left(\frac{y}{y_0}\right) \\ \text{(b) } \lambda = \frac{1}{t}\ln\left(\frac{N_0}{N}\right)\end{array}\right]$$

15 Two quantities x and y are related by the equation $y = a\,e^{-kx}$, where a and k are constants. (a) Determine the value of y when $a = 2.114$, $k = -3.20$ and $x = 1.429$; (b) Determine the value of x when $y = 115.4$, $a = 17.8$ and $k = 4.65$.

[(a) 204.7; (b) –0.4020]

16 The pressure p pascals at height h metres above ground level is given by $p = p_0e^{-\frac{h}{C}}$, where p_0 is the pressure at ground level and C is a constant. When p_0 is 1.012×10^5 Pa and the pressure at a height of 1420 m is 9.921×10^4 Pa, determine the value of C.

[71500]

17 The length l metres of a metal bar at temperature t°C is given by $l = l_0e^{\alpha t}$, where l_0 and α are constants. Determine (a) the value of α when $l = 1.993$ m, $l_0 = 1.894$ m and $t = 250$°C, and (b) the value of l_0 when $l = 2.416$, $t = 310$°C and $\alpha = 1682 \times 10^{-4}$

[(a) 2.038×10^{-4}; (b) 2.293 m]

18 The temperature θ_2 °C of an electrical conductor at time t seconds is given by $\theta_2 = \theta_1 (1 - e^{-\frac{t}{T}})$, where θ_1 is the initial temperature and T seconds is a constant. Determine (a) θ_1 when $\theta_2 = 50$°C, $t = 30$ s and $T = 80$ s, and (b) the time t for θ_2 to fall to half the value of θ_1 if T remains at 80 s.

[(a) 159.9°C; (b) 55.45 s]

19 Quantities x and y are related by $y = 8.317 (1 - e^{\frac{cx}{t}})$, where c and t are constants. Determine (a) the value of y when $c = 2.9 \times 10^{-3}$, $x = 841.2$ and $t = 4.379$, and (b) the value of t when $y = -83.68$, $x = 841.2$ and $c = 2.9 \times 10^{-2}$.

[(a) −6.201; (b) 10.15]

20 The voltage drop, v volts, across an inductor L henrys at time t seconds is given by $v = 200\, e^{-\frac{Rt}{L}}$, where $R = 150$ ohms and $L = 12.5 \times 10^{-3}$ henrys. Determine (a) the voltage when $t = 160 \times 10^{-6}$ s and (b) the time for the voltage to reach 85 volts.

[(a) 29.32 V; (b) 71.31×10^{-6}]

9 The solution of equations

A. MAIN POINTS CONCERNED WITH THE SOLUTION OF EQUATIONS

1 An **equation** is a statement that two quantities are equal. To '**solve an equation**' means 'to find the value of the unknown'. The value of the unknown is called the **root** of the equation.

2 A **linear or simple equation** is one in which an unknown quantity is raised to the power 1. For example, $3x + 2 = 0$ is a linear equation. (See *Problems 1 to 14.*)

3 There are three methods of solving a pair of **linear simultaneous equations** in two unknowns. These are:
(i) by substitution; (ii) by elimination; (iii) graphically.
(See *Problems 15 to 19*)

4 A **quadratic equation** is one in which the highest power of the unknown quantity is 2. For example, $x^2 - 3x + 1 = 0$ is a quadratic equation.

5 There are four methods of solving **quadratic equations**. These are:
(i) by factorisation (where possible),
(ii) by 'completing the square',
(iii) by using the 'quadratic formula', or
(iv) graphically.

6 Multiplying out $(2x + 1)(x - 3)$ gives $2x^2 - 6x + x - 3$, i.e., $2x^2 - 5x - 3$. The reverse process of moving from $2x^2 - 5x - 3$ to $(2x + 1)(x - 3)$ is called **factorising**. If the quadratic expression can be factorised this provides the simplest method of solving a quadratic equation. For example, if $2x^2 - 5x - 3 = 0$, then, by factorising:
$(2x + 1)(x - 3) = 0$
Hence either $(2x + 1) = 0$, i.e., $x = -\dfrac{1}{2}$
or $(x - 3) = 0$, i.e., $x = 3$
The technique of factorising is often one of 'trial and error'.

7 If $ax^2 + bx + c = 0$ then $x = \dfrac{-b \pm \sqrt{(b^2 - 4ac)}}{2a}$
This is known as the **quadratic formula** and is obtained by '**,completing the square**'.
(See *Problems 20 to 30*)

8 There are many practical problems involving simple, simultaneous and quadratic equations to be found in technology and some are demonstrated in the following worked problems.

B. WORKED PROBLEMS ON THE SOLUTION OF EQUATIONS

SIMPLE EQUATIONS

Problem 1 Solve $4(2r - 3) - 2(r - 4) = 3(r - 3) - 1$.

Removing brackets gives: $8r - 12 - 2r + 8 = 3r - 9 - 1$

Rearranging gives: $8r - 2r - 3r = -9 - 1 + 12 - 8$

i.e. $3r = -6$

$$r = \frac{-6}{3} = -2$$

Check: L.H.S. $= 4(-4 - 3) - 2(-2 - 4) = -28 + 12 = -16$.

R.H.S. $= 3(-2 - 3) - 1 = -15 - 1 = -16$.

Hence the solution $r = -2$ is correct.

Problem 2 Solve $\frac{2y}{5} + \frac{3}{4} + 5 = \frac{1}{20} - \frac{3y}{2}$

The L.C.M. of the denominators is 20.

Multiplying each term by 20 gives: $20(\frac{2y}{5}) + 20(\frac{3}{4}) + 20(5) = 20(\frac{1}{20}) - 20(\frac{3y}{2})$

Cancelling gives: $4(2y) + 5(3) + 100 = 1 - 10(3y)$

i.e. $8y + 15 + 100 = 1 - 30y$

Rearranging gives: $8y + 30y = 1 - 15 - 100$

$38y = -114$

$$y = \frac{-114}{38} = -3$$

Check. L.H.S. $= \frac{2(-3)}{5} + \frac{3}{4} + 5 = \frac{-6}{5} + \frac{3}{4} + 5 = \frac{-9}{20} + 5 = 4\frac{11}{20}$.

R.H.S. $= \frac{1}{20} - \frac{3(-3)}{2} = \frac{1}{20} + \frac{9}{2} = 4\frac{11}{20}$

Hence the solution $y = -3$ is correct.

Problem 3 Solve $\frac{3}{t - 2} = \frac{4}{3t + 4}$.

By 'cross-multiplication': $3(3t + 4) = 4(t - 2)$

Removing brackets gives: $9t + 12 = 4t - 8$

Rearranging gives: $9t - 4t = -8 - 12$

i.e. $5t = -20$

$$t = \frac{-20}{5} = -4$$

Check: L.H.S. $= \frac{3}{-4 - 2} = \frac{3}{-6} = -\frac{1}{2}$

R.H.S. $= \frac{4}{3(-4) + 4} = \frac{4}{-12 + 4} = \frac{4}{-8} = -\frac{1}{2}$

Hence the solution $t = -4$ is correct.

Problem 4 Solve $\dfrac{\sqrt{b}+3}{\sqrt{b}} = 2$.

To remove the fraction each term is multiplied by $\sqrt{b}$

$$\text{Hence } \sqrt{b}\left(\frac{\sqrt{b}+3}{\sqrt{b}}\right) = \sqrt{b}(2)$$

Cancelling gives: $\sqrt{b}+3 = 2\sqrt{b}$

Rearranging gives: $3 = 2\sqrt{b} - \sqrt{b} = \sqrt{b}$

Squaring both sides gives: $9 = b$

Check: L.H.S. $= \dfrac{\sqrt{9}+3}{\sqrt{9}} = \dfrac{3+3}{3} = \dfrac{6}{3} = 2 =$ R.H.S.

Problem 5 A copper wire has a length, l, of 1.5 km, a resistance R of 5 Ω and a resistivity ρ of 17.2×10^{-6} Ω mm. Find the cross-sectional areas, a, of the wire, given that $R = \dfrac{\rho l}{a}$.

Since $R = \dfrac{\rho l}{a}$ then $5\ \Omega = \dfrac{(17.2 \times 10^{-6}\ \Omega\ \text{mm})(1500 \times 10^3\ \text{mm})}{a}$

From the units given, a is measured in mm^2.

Thus $5a = 17.2 \times 10^{-6} \times 1500 \times 10^3$

$$a = \frac{17.2 \times 10^{-6} \times 1500 \times 10^3}{5}$$

$$= \frac{17.2 \times 1500 \times 10^3}{10^6 \times 5} = \frac{17.2 \times 15^{3}}{10 \times \not{5}_1} = 5.16$$

Hence the cross-sectional area of the wire is 5.16 mm^2

Problem 6 A rectangular box with square ends has its length 15 cm greater than its breadth and the total length of its edges is 2.04 m. Find the width of the box and its volume.

Let x cm = width = height of box. Then the length of the box is $(x + 15)$ cm. The length of the edges of the box is: $2(4x) + 4(x + 15)$ cm.

$$\text{Hence } 204 = 2(4x) + 4(x+15)$$
$$204 = 8x + 4x + 60$$
$$204 - 60 = 12x$$
$$\text{i.e. } 144 = 12x \text{ and } x = 12 \text{ cm.}$$

Hence the width of the box is 12 cm.

Volume of box = length × width × height = $(x + 15)(x)(x) = (27)(12)(12)$

= **3888 cm^3**.

Problem 7 The temperature coefficient of resistance α may be calculated from the formula $R_t = R_0(1 + \alpha t)$. Find α given $R_t = 0.928$, $R_0 = 0.8$ and $t = 40$.

Since $R_t = R_0(1 + \alpha t)$ then $0.928 = 0.8\,[1 + \alpha(40)]$

$$0.928 = 0.8 + (0.8)(\alpha)(40)$$
$$0.928 - 0.8 = 32\,\alpha$$
$$0.128 = 32\,\alpha$$
$$\text{Hence } \alpha = \frac{0.128}{32} = \mathbf{0.004}$$

Problem 8 The distance s metres travelled in time t seconds is given by the formula $s = ut + \frac{1}{2} a t^2$, where u is the initial velocity in m/s and a is the acceleration in m/s^2. Find the acceleration of the body if it travels 168 m in 6 s, with an initial velocity of 10 m/s.

$s = ut + \frac{1}{2} a t^2$, and $s = 168$, $u = 10$ and $t = 6$.

$$\begin{aligned} \text{Hence } 168 &= (10)(6) + \frac{1}{2} a (6)^2 \\ 168 &= 60 + 18a \\ 168 - 60 &= 18a \\ 108 &= 18a \\ a &= \frac{108}{18} = 6 \end{aligned}$$

Hence the acceleration of the body is 6 m/s²

Problem 9 When three resistors in an electrical circuit are connected in parallel the total resistance R_T is given by $\frac{1}{R_T} = \frac{1}{R_1} + \frac{1}{R_2} + \frac{1}{R_3}$

Find the total resistance when $R_1 = 5$ ohms, $R_2 = 10$ ohms and $R_3 = 30$ ohms.

$$\frac{1}{R_T} = \frac{1}{5} + \frac{1}{10} + \frac{1}{30} = \frac{6 + 3 + 1}{30} = \frac{10}{30} = \frac{1}{3}$$

Taking the reciprocal of both sides gives: R_T **= 3 ohms.**

Alternatively, if $\frac{1}{R_T} = \frac{1}{5} + \frac{1}{10} + \frac{1}{30}$ the L.C.M. of the denominators is $30\,R_T$.

Hence $30\,R_T\left(\frac{1}{R_T}\right) = 30\,R_T\left(\frac{1}{5}\right) + 30\,R_T\left(\frac{1}{10}\right) + 30\,R_T\left(\frac{1}{30}\right)$

Cancelling gives: $30 = 6\,R_T + 3\,R_T + R_T$

i.e. $30 = 10\,R_T$

$R_T = \frac{30}{10} = $ **3 ohms**, as above

Problem 10 The extension x m of an aluminium tie bar of length l m and cross-sectional area A m² when carrying a load of F newtons is given by the modulus of elasticity $E = \frac{Fl}{Ax}$. Find the extension of the tie bar (in mm) if $E = 70 \times 10^9$ N/m², $F = 20 \times 10^6$ N, $A = 0.1$ m² and $l = 1.4$ m.

$E = \frac{Fl}{Ax}$. Hence $70 \times 10^9 \frac{\text{N}}{\text{m}^2} = \frac{(20 \times 10^6 \text{ N})(1.4 \text{ m})}{(0.1 \text{ m}^2)(x)}$

(the unit of x is thus metres).

$$\begin{aligned} 70 \times 10^9 \times 0.1 \times x &= 20 \times 10^6 \times 1.4 \\ x &= \frac{20 \times 10^6 \times 1.4}{70 \times 10^9 \times 0.1} \end{aligned}$$

Cancelling gives $x = \frac{2 \times 1.4}{7 \times 100}$ m $= \frac{2 \times 1.4}{7 \times 100} \times 1000$ mm

Hence the extension of the tie bar, x = 4 mm

Problem 11 Power in a d.c. circuit is given by $P = V^2/R$, where V is the supply voltage and R is the circuit resistance. Find the supply voltage if the circuit resistance is 1.25 ohms and the power measured is 320 W.

Since $P = \frac{V^2}{R}$ then $320 = \frac{V^2}{1.25}$

$$(320)(1.25) = V^2$$

i.e. $V^2 = 400$

Supply voltage, $V = \sqrt{400} = \pm 20$ **V**

Problem 12 A painter is paid £2.25 per hour for a basic 36 hour week, and overtime is paid at one and a third times this rate. Determine how many hours the painter has to work in a week to earn £114.

Basic rate per hour = £2.25; Overtime rate per hour = $1\frac{1}{3} \times$ £2.25 = £3.

Let the number of overtime hours worked = x.

Then $(36)(2.25) + (x)(3) = 114$

$$81 + 3x = 114$$

$$3x = 114 - 81 = 33$$

$$x = \frac{33}{3} = 11$$

Thus 11 hours overtime would have to be worked to earn £114 per week. Hence the total number of hours worked is 36 + 11, i.e. **47 hours**

Problem 13 A formula relating initial and final states of pressures, P_1 and P_2, volumes, V_1 and V_2, and absolute temperatures, T_1 and T_2, of an ideal gas is $\frac{P_1 V_1}{T_1} = \frac{P_2 V_2}{T_2}$

Find the value of P_2 given $P_1 = 100 \times 10^3$, $V_1 = 1.0$, $V_2 = 0.266$, $T_1 = 423$ and $T_2 = 293$.

Since $\frac{P_1 V_1}{T_1} = \frac{P_2 V_2}{T_2}$ then $\frac{(100 \times 10^3)(1.0)}{423} = \frac{P_2\,(0.266)}{293}$

'Cross-multiplying' gives: $(100 \times 10^3)(1.0)(293) = P_2(0.266)(423)$

$$P_2 = \frac{(100 \times 10^3)(1.0)(293)}{(0.266)(423)}$$

Hence, $P_2 = \mathbf{260 \times 10^3}$ **or** $\mathbf{2.6 \times 10^5}$

Problem 14 The stress f in a material of a thick cylinder can be obtained from $\frac{D}{d} = \sqrt{\left(\frac{f+p}{f-p}\right)}$. Calculate the stress given that $D = 21.5$, $d = 10.75$ and $p = 1800$.

Since $\frac{D}{d} = \sqrt{\left(\frac{f+p}{f-p}\right)}$ then $\frac{21.5}{10.75} = \sqrt{\left(\frac{f+1800}{f-1800}\right)}$

i.e. $2 = \sqrt{\left(\frac{f+1800}{f-1800}\right)}$

Squaring both sides gives: $4 = \frac{f+1800}{f-1800}$

$$4(f - 1800) = f + 1800$$
$$4f - 7200 = f + 1800$$
$$4f - f = 1800 + 7200$$
$$3f = 9000$$
$$f = \frac{9000}{3} = 3000$$

Hence the stress, *f*, is 3000

Further problems on simple equations may be found in Section C, Problems 1 to 30, page 120.

SIMULTANEOUS EQUATIONS

Problem 15 Solve, by a substitution method, the simultaneous equations:

$3x - 2y = 12$ (1)

$x + 3y = -7$ (2)

From equation (2), $x = -7 - 3y$.

Substituting for x in equation (1) gives: $3(-7 - 3y) - 2y = 12$

i.e. $-21 - 9y - 2y = 12$

$-11y = 12 + 21 = 33$

Hence $y = \frac{33}{-11} = -3$

Substituting $y = -3$ in equation (2) gives: $x + 3(-3) = -7$

i.e. $x - 9 = -7$

Hence $x = -7 + 9 = 2$

Thus **$x = 2$, $y = -3$** is the solution of the simultaneous equations. (Such solutions should always be checked by substituting values into each of the original two equations.)

Problem 16 Use an elimination method to solve the simultaneous equations

$3x + 4y = 5$ (1)

$2x - 5y = -12$ (2)

If equation (1) is multiplied throughout by 2 and equation (2) by 3, then the coefficient of x will be the same in the newly formed equations. Thus

2 × equation (1) gives: $6x + 8y = 10$ (3)

3 × equation (2) gives: $6x - 15y = -36$ (4)

Equation (3) – equation (4) gives: $0 + 23y = 46$

i.e. $y = \frac{46}{23} = 2$

(Note $+ 8y - -15y = 8y + 15y = 23y$ and $10 - -36 = 10 + 36 = 46$. Alternatively, 'change the signs of the bottom line and add').

Substituting $y = 2$ in equation (1) gives: $3x + 4(2) = 5$,

from which, $3x = 5 - 8 = -3$ and $x = -1$

Checking in equation (2), left hand side $= 2(-1) - 5(2) = -2 - 10 = -12$ = right hand side

Hence **$x = -1$ and $y = 2$** is the solution of the simultaneous equations.

The elimination method is the most common method of solving simultaneous equations.

Problem 17 Solve graphically the equations $1.20x + y = 1.80$
$x - 5.0y = 8.50$

Rearranging each equation into
$y = mx + c$ form gives:

$y = 1.20x + 1.80$ (1)

$y = \frac{x}{5.0} - \frac{8.5}{5.0}$, i.e. $y = 0.20x - 1.70$ (2)

Three co-ordinates are calculated for each equation as shown below.

x	0	1	2
$y = 1.20x + 1.80$	1.80	0.60	–0.60

x	0	1	2
$y = 0.20x - 1.70$	–1.70	–1.50	–1.30

The two lines are plotted as shown in *Fig 1*. The point of intersection is (2.50, –1.20). Hence the solution of the simultaneous equations is x = **2.50**, y = **–1.20**

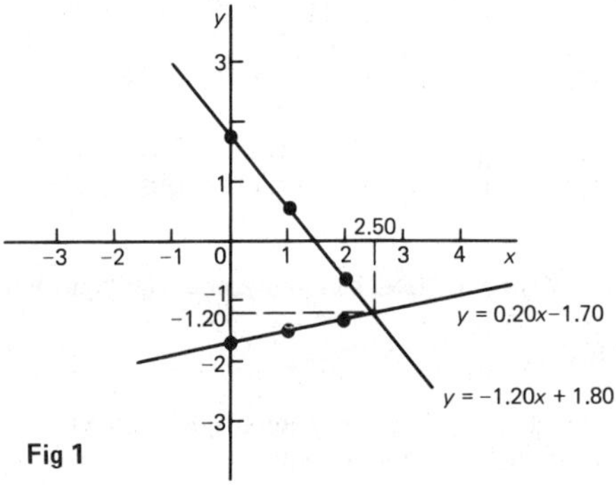

Fig 1

(It is sometimes useful to initially sketch the two straight lines to determine the region where the point of intersection is. Then, for greater accuracy, a graph having a smaller range of values can be drawn to 'magnify' the point of intersection.)

Problem 18 The molar heat capacity of a solid compound is given by the equation $c = a + b\,T$, where a and b are constants. When $c = 52$, $T = 100$ and when $c = 172$, $T = 400$. Determine the values of a and b.

When $c = 52$, $T = 100$. Hence $52 = a + 100b$ (1)
When $c = 172$, $T = 400$. Hence $172 = a + 400b$ (2)
Equation (2) – equation (1) gives: $120 = 300b$,

from which, $b = \frac{120}{300} = 0.4$

Substituting $b = 0.4$ in equation (1) gives: $52 = a + 100\,(0.4)$

$$a = 52 - 40 = 12$$

Hence $a = 12$ and $b = 0.4$

Problem 19 Applying Kirchhoff's laws to an electrical circuit produces the following equations:

$$4.5 = I_1 + 0.5\,(I_1 + I_2) \qquad (1)$$

$$8.5 = 2I_2 + 0.5\,(I_1 + I_2) \qquad (2)$$

Determine the values of currents I_1 and I_2.

Rearranging equation (1) gives: $4.5 = I_1 + 0.5\,I_1 + 0.5\,I_2$

i.e. $1.5\,I_1 + 0.5\,I_2 = 4.5 \qquad (3)$

Rearranging equation (2) gives: $8.5 = 2\,I_2 + 0.5\,I_1 + 0.5\,I_2$

i.e. $0.5\,I_1 + 2.5\,I_2 = 8.5 \qquad (4)$

3 × equation (4) gives: $1.5\,I_1 + 7.5\,I_2 = 25.5 \qquad (5)$

Equation (5) – equation (3) gives: $7.0\,I_2 = 21$

from which $I_2 = \dfrac{21}{7.0} = 3$

Substituting $I_1 = 3$ in equation (3) gives: $1.5\,I_1 + 0.5\,(3) = 4.5$

$$1.5\,I_1 = 4.5 - 1.5 = 3$$

from which, $I_1 = \dfrac{3}{1.5} = 2$

Hence **$I_1 = 2$ and $I_2 = 3$**, which may be checked in the original equations.

Further problems on simultaneous equtions may be found in Section C, Problems 31 to 41, page 122.

QUADRATIC EQUATIONS

Problem 20 The roots of a quadratic equation are $\frac{1}{3}$ and -2. Determine the equation.

If the roots of a quadratic equation are α and β then $(x - \alpha)(x - \beta) = 0$. Hence if $\alpha = \frac{1}{3}$ and $\beta = -2$, then $(x - \frac{1}{3})(x - (-2)) = 0$

$$\text{i.e. } (x - \tfrac{1}{3})(x + 2) = 0$$

$$x^2 - \tfrac{1}{3}x + 2x - \tfrac{2}{3} = 0$$

$$x^2 + \tfrac{5}{3}x - \tfrac{2}{3} = 0$$

$$\text{Hence } \mathbf{3x^2 + 5x - 2 = 0}$$

Problem 21 Find the equations whose roots are
(a) 5 and –5; and
(b) 1.2 and –0.4.

(a) If 5 and –5 are the roots of a quadratic equation then

$(x - 5)(x + 5) = 0$

i.e. $x^2 - 5x + 5x - 25 = 0$

i.e. $\mathbf{x^2 - 25 = 0}$

(b) If 1.2 and −0.4 are the roots of a quadratic equation then

$(x - 1.2)(x + 0.4) = 0$

i.e. $x^2 - 1.2x + 0.4x - 0.48 = 0$

i.e. $\mathbf{x^2 - 0.8x - 0.48 = 0}$

Problem 22 Solve the equations (a) $x^2 + 2x - 8 = 0$ and (b) $3x^2 - 11x - 4 = 0$, by factorisation.

(a) $x^2 + 2x - 8 = 0$

The factors of x^2 are x and x. These are placed in brackets thus:

$(x \quad)(x \quad)$

The factors of −8 are +8 and −1, or −8 and +1, or +4 and −2, or −4 and +2. The only combination to give a middle term of $+2x$ is +4 and −2.

i.e. $x^2 + 2x - 8 = (x + 4)(x - 2)$

(Note that the product of the two inner terms added to the product of the two outer terms must equal the middle terms, $+2x$ in this case.) The quadratic equation $x^2 + 2x - 8 = 0$ thus becomes $(x + 4)(x - 2) = 0$. Since the only way that this can be true is for either the first or the second, or both factors to be zero, then

either $(x + 4) = 0$, i.e., $x = -4$

or $(x - 2) = 0$, i.e., $x = 2$

Hence the roots of $x^2 + 2x - 8 = 0$ are $x = -4$ and 2

(b) $3x^2 - 11x - 4 = 0$

The factors of $3x^2$ are $3x$ and x. These are placed in brackets thus:

$(3x \quad)(x \quad)$

The factors of −4 are −4 and +1, or +4 and −1, or −2 and 2. Remembering that the product of the two inner terms added to the product of the two outer terms must equal $-11x$, the only combination to give this is −4 and +1.

i.e. $3x^2 - 11x - 4 = (3x + 1)(x - 4)$.

The quadratic equation $3x^2 - 11x - 4 = 0$ thus becomes $(3x + 1)(x - 4) = 0$.

Hence, either $(3x + 1) = 0$, i.e., $x = -\frac{1}{3}$

or $(x - 4) = 0$, i.e., $x = 4$, and both solutions may be checked in the original equation.

Problem 23 Determine the roots of (a) $x^2 - 6x + 9 = 0$ and (b) $4x^2 - 25 = 0$, by factorisation.

(a) $x^2 - 6x + 9 = 0$. Hence $(x - 3)(x - 3) = 0$.

i.e. $(x - 3)^2 = 0$, (the left hand side is known as a **perfect square**).

Hence $x = 3$ is the only root of the equation $x^2 - 6x + 9 = 0$.

(b) $4x^2 - 25 = 0$ (the left hand side is the difference of two squares, $(2x)^2$ and $(5)^2$

Hence $(2x + 5)(2x - 5) = 0$.

Hence either $(2x + 5) = 0$, i.e., $x = -\frac{5}{2}$

or $(2x - 5) = 0$, i.e., $x = \frac{5}{2}$.

Problem 24 Solve $2x^2 + 9x + 8 = 0$, correct to 3 significant figures, by completing the square.

Making the coefficient of x^2 unity gives: $x^2 + \frac{9}{2}x + 4 = 0$

and rearranging gives: $x^2 + \frac{9}{2}x = -4.$

Adding to both sides (half the coefficient of $x)^2$ gives: $x^2 + \frac{9}{2}x + \left(\frac{9}{4}\right)^2 = \left(\frac{9}{4}\right)^2 - 4$

The L.H.S. is now a perfect square, thus $\left(x + \frac{9}{4}\right)^2 = \frac{81}{16} - 4 = \frac{17}{16}$

Taking the square root of both sides gives: $x + \frac{9}{4} = \sqrt{\frac{17}{16}} = \pm 1.031$

Hence $x = -\frac{9}{4} \pm 1.031,$

i.e. x = **−3.28 or −1.22**, correct to 3 significant figures.

Problem 25 Solve $4x^2 + 7x + 2 = 0$, giving the roots correct to 2 decimal places.

Comparing $4x^2 + 7x + 2 = 0$ with $ax^2 + bx + c = 0$ gives $a = 4$, $b = 7$ and $c = 2$.

Hence $x = \frac{-7 \pm \sqrt{[(7)^2 - 4(4)(2)]}}{2(4)} = \frac{-7 \pm \sqrt{17}}{8} = \frac{-7 \pm 4.123}{8} = \frac{-7 + 4.123}{8}$

or $\frac{-7 - 4.123}{8}$

Hence x = **−0.36 or −1.39**, correct to 2 decimal places.

Problem 26 Solve graphically the quadratic equation $-5x^2 + 9x + 7.2 = 0$, given that the solutions lie between $x = -1$ and $x = 3$. Determine also the co-ordinates of the turning point and state its nature.

Let $y = -5x^2 + 9x + 7.2$. A table of values is drawn up as shown in Table 1.

Table 1

x	−1	−0.5	0	1	2	2.5	3
$-5x^2$	−5	−1.25	0	−5	−20	−31.25	−45
$+9x$	−9	−4.5	0	9	18	22.5	27
$+7.2$	7.2	7.2	7.2	7.2	7.2	7.2	7.2
$y = -5x^2 + 9x + 7.2$	−6.8	1.45	7.2	11.2	5.2	−1.55	−10.8

A graph of $y = -5x^2 + 9x + 7.2$ is shown plotted in *Fig 2*. The graph crosses the x-axis (i.e. where $y = 0$) at x = **−0.6** and x = **2.4** and these are the solutions of the quadratic equation $-5x^2 + 9x + 7.2 = 0$. The turning point is a **maximum** having co-ordinates **(0.9, 11.25).**

Problem 27. Plot a graph of $y = 2x^2$ and hence solve the equations (a) $2x^2 - 8 = 0$; and (b) $2x^2 - x - 3 = 0$

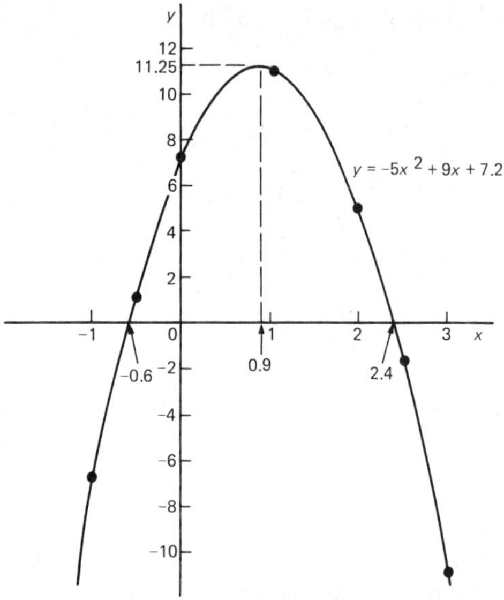

Fig 2

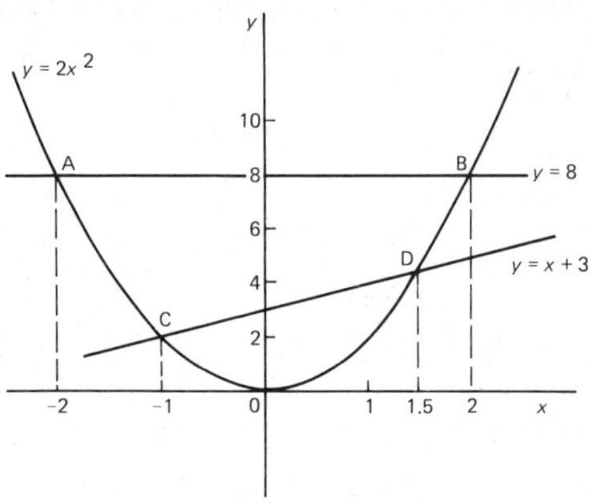

Fig 3

A graph of $y = 2x^2$ is shown in *Fig 3*.
(a) Rearranging $2x^2 - 8 = 0$ gives $2x^2 = 8$ and the solution of this equation is obtained from the points of intersection of $y = 2x^2$ and $y = 8$, i.e., at co-ordinates $(-2, 8)$ and $(2, 8)$, shown as A and B respectively in fig 3. Hence the solutions of $2x^2 - 8 = 0$ are $\mathbf{x = -2}$ **and** $\mathbf{x = +2}$.
(b) Rearranging $2x^2 - x - 3 = 0$ gives $2x^2 = x + 3$ and the solution of this equation is obtained from the points of interesection of $y = 2x^2$ and $y = x + 3$, i.e., at C and D in *Fig 3*. Hence the solutions of $2x^2 - x - 3 = 0$ are $\mathbf{x = -1}$ **and** $\mathbf{x = 1.5}$

Problem 28 The area of a rectangle is 23.6 cm^2 and its width is 3.10 cm shorter than its length. Determine the dimensions of the rectangle, correct to 3 significant figures.

Let the length of the rectangle be x cm. Then the width is $(x - 3.10)$ cm.
Area = length × width = $x(x - 3.10) = 23.6$

i.e., $x^2 - 3.10x - 23.6 = 0$

Using the quadratic formula, $x = \dfrac{-(-3.10) \pm \sqrt{[(-3.10)^2 - 4(1)(-23.6)]}}{2(1)}$

$= \dfrac{3.10 \pm \sqrt{(9.61 + 94.4)}}{2} = \dfrac{3.10 \pm 10.20}{2}$

$= \dfrac{13.30}{2}$ or $\dfrac{-7.10}{2}$

Hence $x = 6.65$ cm or -3.55 cm. The latter solution is neglected since length cannot be negative.
Thus length $x = 6.65$ cm and width $= x - 3.10 = 6.65 - 3.10 = 3.55$ cm.
Hence the dimensions of the rectangle are 6.61 cm by 3.55 cm.
(*Check:* Area = 6.65 × 3.55 = 23.6 cm^2, correct to 3 significant figures.)

Problem 29 Calculate the diameter of a solid cylinder which has a height of 82.0 cm and a total surface area of 2.0 m^2.

Total surface area of a cylinder = curved surface area + 2 circular ends
$= 2\pi rh + 2\pi r^2$ (where r = radius and h = height)

Since the total surface area = 2.0 m^2 and the height h = 82 cm or 0.82 m,
then $2.0 = 2\pi r(0.82) + 2\pi r^2$
i.e. $2\pi r^2 + 2\pi r(0.82) - 2.0 = 0$

Dividing throughout by 2π gives: $r^2 + 0.82\,r - \dfrac{1}{\pi} = 0$

Using the quadratic formula: $r = \dfrac{-0.82 \pm \sqrt{[(0.82)^2 - 4(1)(-\frac{1}{\pi})]}}{2(1)}$

$= \dfrac{-0.82 \pm \sqrt{1.9456}}{2} = \dfrac{-0.82 \pm 1.3948}{2}$

$= 0.2874$ or -1.1074

Thus the radius r of the cylinder is 0.2874 m (the negative solution being neglected.)
Hence the diameter of the cylinder = 2 × 0.2874
= **0.5748 m or 57.5 cm**, correct to 3 significant figures.

Problem 30 The height s metres of a mass projected vertically upwards at time t seconds is $s = ut - \frac{1}{2}g\,t^2$. Determine how long the mass will take after being projected to reach a height of 16 m (a) on the ascent and (b) on the descent, when $u = 30$ m/s and $g = 9.81$ m/s^2.

When height $s = 16$ m, $16 = 30t - \frac{1}{2}(9.81)\,t^2$

$$\text{i.e. } 4.905\,t^2 - 30t + 16 = 0$$

Using the quadratic formula: $t = \dfrac{-(-30) \pm \sqrt{[(-30)^2 - 4(4.905)(16)]}}{2(4.905)}$

$$= \frac{30 \pm \sqrt{586.1}}{9.81} = \frac{30 \pm 24.21}{9.81} = 5.53 \text{ or } 0.59$$

Hence the mass will reach a height of 16 m after 0.59 s on the ascent and after 5.53 s on the descent.

Further problems on quadratic equations may be found in the following Section C, Problems 42 to 58, page 120.

C. FURTHER PROBLEMS ON THE SOLUTION OF EQUATIONS

SIMPLE EQUATIONS

Solve the following equations:

1 $\frac{1}{5}(2f - 3) + \frac{1}{6}(f - 4) + \frac{2}{15} = 0.$ [2]

2 $\frac{1}{3}(3m - 6) - \frac{1}{4}(5m + 4) + \frac{1}{5}(2m - 9) = -3$ [12]

3 $\frac{x}{3} - \frac{x}{5} = 2.$ [15]

4 $1 - \frac{y}{3} = 3 + \frac{y}{3} - \frac{y}{6}$ [−4]

5 $\frac{3t}{20} = \frac{6 - t}{12} + \frac{2t}{15} - \frac{3}{2}$ [−10]

6 $\frac{y}{5} + \frac{7}{20} = \frac{5 - y}{4}$ [2]

7 $\frac{v - 2}{2v - 3} = \frac{1}{3}$ [3]

8 $\frac{2}{a - 3} = \frac{3}{2a + 1}$ [−11]

9 $\frac{2c - 3}{4} - \frac{1 - c}{5} - 1 = \frac{2c + 3}{3} + \frac{43}{60}$ [110]

10 $4 = \sqrt{\left(\frac{3}{a}\right)} + 3$ [3]

11 $\frac{3\sqrt{x}}{1 - \sqrt{x}} = -6$ [4]

12 $10 = 5\sqrt{\left(\frac{x}{2} - 1\right)}$ [10]

13 $16 = \frac{t^2}{9}$ [±12]

14 $\sqrt{\left(\frac{y + 2}{y - 2}\right)} = \frac{1}{2}$ $[-3\frac{1}{3}]$

15 $\frac{6}{a} = \frac{2a}{3}$ [±3]

16 $\frac{11}{2} = 5 + \frac{8}{x^2}$ [±4]

17 A formula used for calculating resistance of a cable is $R = \frac{\rho l}{a}$.

Given $R = 1.25$, $l = 2500$ and $a = 2 \times 10^{-4}$, find the value of ρ.

[10^{-7}]

18 Force F newtons is given by $F = m\,a$, where m is the mass in kilograms and a is the acceleration in metres per second squared. Find the acceleration when a force of 4 kN is applied to a mass of 500 kg.

[8 m/s²]

19 $PV = mRT$ is the characteristic gas equation. Find the value of m when $P = 100 \times 10^3$, $V = 3.00$, $R = 288$ and $T = 300$.

[3.472]

20 When three resistors R_1, R_2 and R_3 are connected in parallel the total resistance R_T is determined from $\frac{1}{R_T} = \frac{1}{R_1} + \frac{1}{R_2} + \frac{1}{R_3}$.

(a) Find the total resistance when $R_1 = 3\Omega$, $R_2 = 6\Omega$ and $R_3 = 18\Omega$.

(b) Find the value of R_3 given that $R_T = 3\Omega$, $R_1 = 5\Omega$ and $R_2 = 10\Omega$.

[(a) 1.8Ω (b) 30Ω]

21 Five pens and two rulers cost 94p. If a ruler costs 5p more than a pen, find the cost of each.

[12p, 17p]

22 Ohm's law may be represented by $I = \frac{V}{R}$, where I is the current in amperes, V is the voltage in volts, and R is the resistance in ohms. A soldering iron takes a current of 0.30 A from a 240 V supply. Find the resistance of the element.

[800Ω]

23 A rectangle has a length of 20 cm and a width b cm. When its width is reduced by 4 cm its area becomes 160 cm². Find the original width and area of the rectangle.

[12 cm, 240 cm²]

24 Given $R_2 = R_1(1 + \alpha t)$, find α given $R_1 = 5.0$, $R_2 = 6.03$ and $t = 51.5$.

[0.004]

25 If $v^2 = u^2 + 2\,as$, find u given $v = 24$, $a = -40$ and $s = 4.05$.

[30]

26 The relationship between the temperature on a Fahrenheit scale and that on a Celsius scale is given by $F = \frac{9}{5}C + 32$. Express 113°F in degrees Celsius.

[45°C]

27 If $t = 2\pi\sqrt{\left(\frac{w}{S\,g}\right)}$, find the value of S given $w = 1.219$, $g = 9.81$ and $t = 0.313\,2$.

[50]

28 Two joiners and five mates earn £484 between them for a particular job. If a joiner earns £18 more than a mate, calculate the earnings for a joiner and for a mate.

[£82, £64]

29 An alloy contains 60% by weight of copper, the remainder being zinc. How much copper must be mixed with 50 kg of this alloy to give an alloy containing 75% copper? [30 kg]

30 A rectangular laboratory has a length equal to one and a half times its width and a perimeter of 40 m. Find its length and width.

[12 m, 8 m]

SIMULTANEOUS EQUATIONS

In *Problems 31 to 38*, solve the given simultaneous equations by elimination, substitution or graphical methods, and verify the results.

31 $3x + 2y = 12$; $4x - y = 5$

[$x = 2; y = 3$]

32 $5x - 3y = 11$; $3x + y = 8$

[$x = 2\frac{1}{2}; y = \frac{1}{2}$]

33 $5x = 2y$; $3x = 41 - 7y$

[$x = 2; y = 5$]

34 $\frac{x}{2} - 7 + 2y = 0$; $5x + \frac{2}{3}y = 12$

[$x = 2; y = 3$]

35 $3y - 2.5x = 0.45$; $1.6x + 0.8y = 0.8$.

[$x = 0.30; y = 0.40$]

36 $\frac{4}{x} - \frac{3}{y} = 18$; $\frac{2}{x} + \frac{5}{y} = -4$

[$x = \frac{1}{3}; y = -\frac{1}{2}$]

37 $\frac{x+1}{4} + 1 - \frac{y+2}{3} = 0$; $\frac{1-x}{5} + \frac{3-y}{4} + \frac{13}{20} = 0$

[$x = 3; y = 4$]

38 $\frac{x}{4} + \frac{2+y}{3} = \frac{1}{2}$; $\frac{5+x}{6} + \frac{2y}{5} = \frac{9}{10}$

[$x = -2, y = 1$]

39 In a simple machine, the effort E newtons required to overcome a load L newtons is given by: $E = a + b\,L$, where a and b are constants. An effort of 4.2 N overcomes a load of 5 N and an effort of 5 N overcomes a load of 6 N. Determine graphically the values of a and b.

[$a = 0.2; b = 0.8$]

40 Equations produced by applying Kirchhoff's laws to an electrical circuit are:

$4 = 0.3\,I_1 + 1.5\,(I_1 - I_2)$

$10 = 1.5\,(I_2 - I_1) + 0.2\,I_2$.

Determine currents I_1 and I_2, correct to 2 decimal places.

[$I_1 = 26.91; I_2 = 29.63$]

41 The distance s travelled by a vehicle in t seconds is given by: $s = ut + \frac{1}{2}at^2$, where u and a are constants. Determine the values of u and a given that $s = 34$ m when $t = 2$s, and $s = 167.5$ m when $t = 5$s.

[$u = 6; a = 11$]

QUADRATIC EQUATIONS

42 Determine the quadratic equations in x whose roots are (a) 3 and 1; (b) 2 and –5; (c) –1 and –4.

[(a) $x^2 - 4x + 3 = 0$; (b) $x^2 + 3x - 10 = 0$;
(c) $x^2 + 5x + 4 = 0$]

43 Determine the quadratic equations in x whose roots are (a) $2\frac{1}{2}$ and $-\frac{1}{2}$; (b) 6 and -6; (c) 2.4 and -0.7

[(a) $4x^2 - 8x - 5 = 0$; (b) $x^2 - 36 = 0$; (c) $x^2 - 1.7x - 1.68 = 0$.]

In *Problems 44 and 45*, solve the given equations by factorisation.

44 (a) $x^2 + 4x - 32 = 0$; (b) $x^2 - 16 = 0$; (c) $(x + 2)^2 = 16$.

[(a) 4, -8; (b) 4, -4; (c) 2, -6]

45 (a) $2x^2 - x - 3 = 0$; (b) $6x^2 - 5x + 1 = 0$; (c) $10x^2 + 3x - 4 = 0$.

$\left[(a) -1, 1\frac{1}{2}; (b) \frac{1}{2}, \frac{1}{3}; (c) \frac{1}{2}, -\frac{4}{5}\right]$

In *Problems 46 and 47*, solve the given equations by using the quadratic formula or by completing the square, correct to 3 decimal places.

46 (a) $x^2 + 4x + 1 = 0$; (b) $2x^2 + 5x - 4 = 0$; (c) $3x^2 - x - 5 = 0$.

[(a) -3.732, -0.268; (b) -3.137, 0.637; (c) 1.468, -1.135]

47 (a) $5x^2 - 8x + 2 = 0$; (b) $4x^2 - 11x + 3 = 0$; (c) $2x^2 + 5x = 2$.

[(a) 1.290, 0.310; (b) 2.443, 0.307; (c) -2.851, 0.351]

48 Solve the quadratic equation $2x^2 + 7x + 6 = 0$ graphically, given that the solutions lie in the range $x = -2$ to $x = 3$. Determine also the nature and co-ordinates of its turning point.

$\left[x = -1\frac{1}{2} \text{ or } 2; \text{ Minimum at } (-1\frac{3}{4}, -\frac{1}{8}\right]$

49 Solve graphically the quadratic equation $10x^2 - 9x - 11.2 = 0$, given that the roots lie between $x = -1$ and $x = 2$.

[$x = -0.7$ or 1.6]

50 Plot a graph of $y = 3x^2$ and hence solve the equations (a) $3x^2 - 8 = 0$ and (b) $3x^2 - 2x - 1 = 0$.

$\left[(a) \pm 1.63; (b) 1 \text{ or } -\frac{1}{3}\right]$

51 The angle a rotating shaft turns through in t seconds is given by $\theta = wt + \frac{1}{2}\alpha t^2$. Determine the time taken to complete 4 radians if w is 3.0 rad/s and α is 0.60 rad/s^2.

[1.191 s]

52 The power P developed in an electrical circuit is given by $P = 10\,I - 8\,I^2$, where I is the current in amperes. Determine the current necessary to produce a power of 2.5 watts in the circuit.

[0.345 A or 0.905 A]

53 The area of a triangle is 47.6 cm^2 and its perpendicular height is 4.3 cm more than its base length. Determine the length of the base correct to 3 significant figures.

[7.84 cm]

54 The sag l metres in a cable stretched between two supports, distance x m apart is given by: $l = \frac{12}{x} + x$. Determine the distance between supports when the sag is 20 m.

[0.619 m or 19.38 m]

55 The acid dissociation constant K_a of ethanoic acid is 1.8×10^{-5} mol dm^{-3} for a particular solution. Using the Ostwald dilution law $K_a = \frac{x^2}{v(1-x)}$ determine x, the degree of ionisation, given that $v = 10$ dm^3.

[0.0133]

56 A rectangular building is 15 m long by 11m wide. A concrete path of constant width is laid all the way around the building. If the area of the path is 60.0 m^2, calculate its width correct to the nearest millimetre.

[1.066 m]

57 The total surface area of a closed cylindrical container is 20.0 m^2. Calculate the radius of the cylinder if its height is 2.80 m.

[86.78 cm]

58 The bending moment M at a point in a beam is given by $M = \dfrac{3x(20 - x)}{2}$, where x metres is the distance from the point of support. Determine the value of x when the bending moment is 50 N m.

[1.835 m or 18.165 m]

10 Probability

A. MAIN POINTS CONCERNED WITH PROBABILITY

1 The **probability** of something happening is the likelihood or chance of it happening. Values of probability lie between 0 and 1, where 0 represents an absolute impossibility and 1 represents an absolute certainty. The probability of an event happening usually lies somewhere between these two extreme values and is expressed either as a proper or a decimal fraction. Examples of probability are:

that a length of copper wire has zero resistance at 100°C	0
that a fair, six-sided dice will stop with a 3 upwards	$\frac{1}{6}$ or 0.1667
that a fair coin will land with a head upwards	$\frac{1}{2}$ or 0.5
that a length of copper wire has some resistance at 100°C	1.

2 If p is the probability of an event happening and q is the probability of the same event not happening, then the total probability is $p+q$ and is equal to unity, since it is an absolute certainty that the event either does or does not occur, i.e., $p+q = 1$.

3 The **expectation,** E, of an event happening is defined in general terms as the product of the probability p of an event happening and the number of attempts made, n, i.e., $E = pn$. Thus, since the probability of obtaining a 3 upwards when rolling a fair dice is 1/6, the expectation of getting a 3 upwards on four throws of the dice is

$\frac{1}{6} \times 4$, i.e., $\frac{2}{3}$.

Thus expectation is the average occurrence of an event.

4 A **dependent event** is one in which the probability of one event happening affects the probability of another event happening. Let 5 transistors be taken at random from a batch of 100 transistors for test purposes, and the probability of there being a defective transistor, p_1, be determined. At some later time, let another 5 transistors be taken at random from the 95 remaining transistors in the batch and the probability of there being a defective transistor, p_2, be determined. The value of p_2 is different from p_1 since the batch size has effectively altered from 100 to 95, i.e., the probability p_2 is dependent on probability p_1. Since 5 transistors are drawn, and then another 5 transistors are drawn without replacing the first 5, the second random selection is said to be **without replacement.**

5 An **independent event** is one in which the probability of an event happening does **not** affect the probability of another event happening. If 5 transistors are taken at random from a batch of transistors and the probability of a defective transistor p_1

is determined and the process is repeated after the original 5 have been replaced in the batch to give p_2, then p_1 is equal to p_2. Since the 5 transistors are replaced between draws, the second selection is said to be **with replacement**.

6 **The addition law of probability**
The addition law of probability is recognised by the word '**or**' joining the probabilities. If p_A is the probability of event A happening and p_B is the probability of event B happening, the probability of event A **or** event B happening is given by $p_A + p_B$. Similarly, the probability of events A **or** B **or** C **or** .... N happening is given by

$$p_A + p_B + p_C + \ldots\ldots\ldots + p_N.$$

7 **The multiplication law of probability**
The multiplication law of probability is recognised by the word '**and**' joining the probabilities. If p_A is the probability of event A happening and p_B is the probability of event B happening, the probability of event A **and** event B happening is given by $p_A \times p_B$. Similarly, the probability of events A **and** B **and** C **and** ..... N happening is given by $p_A \times p_B \times p_C \times \ldots\ldots\ldots \times p_N$.

8 The addition and multiplication laws of probability may be combined as shown below. Let p_A, p_B and p_C be the probabilities of events A, B and C respectively happening. The probabilities of events A, B and C **not** happening may be shown as $\bar{p}_A$, $\bar{p}_B$ and $\bar{p}_C$, (where $p_A + \bar{p}_A = 1$, see para. 2). Then, for example:

(i) the probability of events (A **and** B) **or** C happening is $(p_A \times p_B) + p_C$,

(ii) the probability of (event A **or** event B happening) **and** event C happening is $(p_A + p_B) \times p_C$,

(iii) the probability of (events A **and** B happening) **or** (event A happening **and** event C not happening) is $(p_A \times p_B) + (p_A \times \bar{p}_C)$,

(iv) the probability of (events A **and** B not happening) **or** (event C happening) is $(\bar{p}_A \times \bar{p}_B) + p_C$, and so on.

9 When two die (die is the plural of dice) are thrown, the probabilities of scoring 2, 3, 4 ......, 11 or 12 can be determined by considering the matrix of all possible numbers obtainable, as shown. (For example, the score of 7 shown in row 1 is made up of a 6 on dice A and a 1 on dice B).

		Dice A					
		1	2	3	4	5	6
	1	2	3	4	5	6	7
	2	3	4	5	6	7	8
	3	4	5	6	7	8	9
Dice B	4	5	6	7	8	9	10
	5	6	7	8	9	10	11
	6	7	8	9	10	11	12

It can be seen from the matrix that there are 36 possibilities or **sample spaces**. This agrees with the theory of probability that the probability of obtaining a particular number on dice A is $\frac{1}{6}$ and that of obtaining a particular number on dice B is $\frac{1}{6}$, hence the probability of obtaining a given number with both dice A **and** dice B is $\frac{1}{6} \times \frac{1}{6} = \frac{1}{36}$.

From the matrix of sample spaces, the probability of obtaining any particular score between 2 and 12 is the ratio $\frac{\text{sample spaces containing that score}}{\text{total number of sample spaces}}$, giving:

Score from 2 die	2	3	4	5	6	7	8	9	10	11	12
Probability	$\frac{1}{36}$	$\frac{2}{36}$	$\frac{3}{36}$	$\frac{4}{36}$	$\frac{5}{36}$	$\frac{6}{36}$	$\frac{5}{36}$	$\frac{4}{36}$	$\frac{3}{36}$	$\frac{2}{36}$	$\frac{1}{36}$

These results are shown as a symmetrical histogram in *Fig 1(a)*.

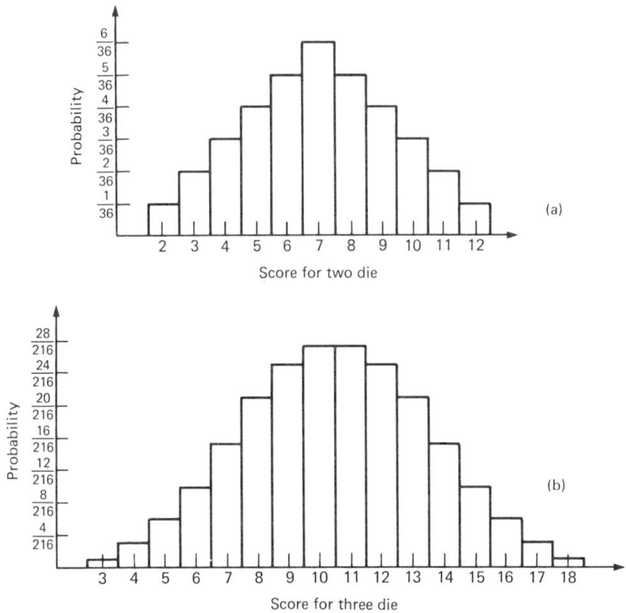

Fig 1

10 When three die are thrown, the probabilities of obtaining scores of 3, 4, 5, ..., 17 or 18 can be obtained by considering the sample spaces for three die. Part of such a matrix is as shown. The 36 possibilities for the scores of die A and B, (see para. 9), are shown horizontally in the top row, the six possibilities for dice C are shown vertically giving 216 sample spaces.

		Die A and B										
		2	3	4	5	6	7	3	4	5	10	11	12
	1	3	4	5	6	7	8	4	5	6	11	12	13
	2	4	5	6	7	8	9	5	6	7	12	13	14
	3	5	6	7	8	9	10	6	7	8	13	14	15
Dice C	4	6	7	8	9	10	11	7	8	9	14	15	16
	5	7	8	9	10	11	12	8	9	10	15	16	17
	6	8	9	10	11	12	13	9	10	11	16	17	18

The probabilities resulting from an analysis of this matrix are:

Score from 3 die	3	4	5	6	7	8	9	10	11	12	13	14	15	16	17	18
Probability	$\frac{1}{216}$	$\frac{3}{216}$	$\frac{6}{216}$	$\frac{10}{216}$	$\frac{15}{216}$	$\frac{21}{216}$	$\frac{25}{216}$	$\frac{27}{216}$	$\frac{27}{216}$	$\frac{25}{216}$	$\frac{21}{216}$	$\frac{15}{216}$	$\frac{10}{216}$	$\frac{6}{216}$	$\frac{3}{216}$	$\frac{1}{216}$

These results are shown as the symmetrical histogram in *Fig 1(b)*.

11 When four die are thrown, the probabilities of obtaining socres of 4, 5, 6, ... , 23 or 24 can be obtained by considering the matrix having 1296 sample spaces. This matrix has the 216 possibilities obtained in para. 10 for three die, as the horizontal top row variables of the matrix together with the six possibilities for dice D as the vertical variables. A histogram of the probabilities is similar in shape to the histogram shown in *Fig 1(b)*, but has 21 rectangles instead of 16. In general, as the number of die being thrown increases, more and more rectangles are introduced and the profile of the histogram approaches more closely to that of a smooth curve.

B. WORKED PROBLEMS ON PROBABILITY

Problem 1 Determine the probabilities of selecting at random (a) a man, and (b) a woman from a crowd containing 20 men and 33 women.

(a) The probability of selecting at random a man, p, is given by the ratio

$$\frac{\text{number of men}}{\text{number in crowd}}, \text{ i.e., } p = \frac{20}{20+33} = \frac{20}{53}.$$

(b) The probability of selecting at random a woman, q, is given by the ratio

$$\frac{\text{number of women}}{\text{number in crowd}}, \text{ i.e., } q = \frac{33}{20+33} = \frac{33}{53}.$$

[Check: the total probability should be equal to 1. $p = \frac{20}{53}$ and $q = \frac{33}{53}$.

The total probability, $p+q = \frac{20}{53}+\frac{33}{53} = 1$, hence no obvious error has been made.]

Problem 2 Find the expectation of obtaining a 4 upwards with 3 throws of a fair dice.

Expectation is the average occurrence of an event and is defined as the probability times the number of attempts, (see para. 3). The probability, p, of obtaining a 4 upwards for one throw of the dice is 1/6. Also, 3 attempts, n, are made, hence the expectation, E, is pn, i.e. $\frac{1}{6} \times 3 = \frac{1}{2}$.

Problem 3 Calculate the probabilities of selecting at random:
(a) the winning horse in a race in which 10 horses are running,
(b) the winning horse in the first race or the winning horse in the second race if there are 10 horses in each race, and
(c) the winning horses in both the first and second races if there are 10 horses in each race.

(a) Since only one of the ten horses can win, the probability of selecting at random the winning horse is
$\frac{\text{number of winners}}{\text{number of horses}}$, i.e. $\frac{1}{10}$.

(b) The probability of selecting at random the winning horse in the first race **or** the winning horse in the second race is given by the addition law of probability, (see para. 6), since the word **or** joins the two probabilities. From (a) above, the probability of selecting the winning horse is 1/10 for the first race and also 1/10 for the second race. Hence, the probability of selecting the winning horse from the first **or** the second race is
$\frac{1}{10} + \frac{1}{10} = \frac{1}{10} = \frac{1}{5}$ **or 0.2**

(c) The probability of selecting the winning horse in the first race is $\frac{1}{10}$.
The probability of selecting the winning horse in the second race is $\frac{1}{10}$.
The probability of selecting the winning horses in the first **and** second race is given by the multiplication law of probability, (see para. 7),
i.e., probability $= \frac{1}{10} \times \frac{1}{10} = \frac{1}{100}$ **or 0.01**

Problem 4 The probability of a component failing in one year due to excessive temperature is 1/20, due to excessive vibration is 1/25 and due to excessive humidity is 1/50. Determine the probabilities that during a one-year period, a component:
(a) fails due to excessive temperature and excessive vibration,
(b) fails due to excessive vibration or excessive humidity and
(c) will not fail because of both excessive temperature and excessive humidity.

Let p_A be the probability of failure due to excessive temperature, then
$p_A = \frac{1}{20}$ and $\bar{p}_A = \frac{19}{20}$.

Let p_B be the probability of failure due to excessive vibration, then
$p_B = \frac{1}{25}$ and $\bar{p}_B = \frac{24}{25}$.

Let p_C be the probability of failure due to excessive humidity, then
$p_C = \frac{1}{50}$ and $\bar{p}_C = \frac{49}{50}$.

(a) The probability of a component failing due to excessive temperature **and** excessive vibration is given by
$p_A \times p_B$, i.e., $\frac{1}{20} \times \frac{1}{25} = \frac{1}{500}$ **or 0.002**

(b) The probability of a component failing due to excessive vibration or excessive humidity is
$p_B + p_C$, i.e., $\frac{1}{25} + \frac{1}{50} = \frac{3}{50}$ **or 0.06**

(c) The probability that a component will not fail due to excessive temperature and will not fail due to excessive humidity is

$\bar{p}_A \times \bar{p}_C$, i.e. $\frac{19}{20} \times \frac{49}{50} = \frac{931}{1000}$ or **0.931**

Problem 5 A batch of 100 capacitors contains 73 which are within the required tolerance values, 17 which are below the required tolerance values and the remainder are above the required tolerance values. Determine the probabilities that when randomly selecting a capacitor and then a second capacitor:
(a) both are within the required tolerance values when selecting with replacement and (b) the first one drawn is below and the second one drawn is above the required tolerance value, when selection is without replacement.

(a) The probability of selecting a capacitor within the required tolerance values is 73/100. The first capacitor drawn is now replaced and a second one is drawn from the batch of 100. The probability of this capacitor being within the required tolerance values is also 73/100. Thus, the probability of selecting a capacitor within the required tolerance values for both the first **and** the second draw is

$$\frac{73}{100} \times \frac{73}{100} = \frac{5329}{10\,000} \text{ or } 0.5329$$

(b) The probability of obtaining a capacitor below the required tolerance values on the first draw is 17/100. There are now only 99 capacitors left in the batch, since the first capacitor is not replaced. The probability of drawing a capacitor above the required tolerance values on the second draw is 10/99, since there are (100–73–17), i.e. 10 capacitors above the required tolerance value. Thus, the probability of randomly selecting a capacitor below the required tolerance values and followed by randomly selecting a capacitor above the tolerance values is

$$\frac{17}{100} \times \frac{10}{99} = \frac{170}{9900} = \frac{17}{990} = 0.0172$$

Problem 6 A batch of 40 components contains 5 which are defective. If a component is drawn at random from the batch and tested and then a second component is drawn, determine the probability that neither of the components is defective.

With replacement:
The probability that the component selected on the first draw is satisfactory is 35/40, i.e. 7/8. The component is now replaced and a second draw is made. The probability that this component is also satisfactory is 7/8. Hence, the probability that both the first component drawn and the second component drawn are satisfactory is

$$\frac{7}{8} \times \frac{7}{8} = \frac{49}{64} \text{ or } 0.7656$$

Without replacement:
The probability that the first component drawn is satisfactory is 7/8. There are now only 34 satisfactory components left in the batch and the batch number is 39. Hence, the probability of drawing a satisfactory component on the second draw is 34/39. Thus the probability that the first component drawn and the second component drawn are satisfactory is

$$\frac{7}{8} \times \frac{34}{39} = \frac{238}{312} \text{ or } 0.7628$$

Problem 7 A batch of 40 components contains 5 which are defective. If a component is drawn at random from the batch and tested and then a second component is drawn at random, calculate the probability of having one defective component, both with and without replacement.

The probability of having one defective component can be achieved in two ways. If p is the probability of drawing a defective component and q is the probability of drawing a satisfactory component, then the probability of having one defective component is given by drawing a satisfactory component and then a defective component or by drawing a defective component and then a satisfactory one, i.e., by $q \times p + p \times q$.

With replacement:

$$p = \frac{5}{40} = \frac{1}{8} \text{ and } q = \frac{35}{40} = \frac{7}{8}.$$

Hence, probability of having a defective component is $\frac{1}{8} \times \frac{7}{8} + \frac{7}{8} \times \frac{1}{8}$,

$$\text{i.e. } \frac{7}{64} + \frac{7}{64} = \frac{14}{64} = \frac{7}{32} \text{ or } 0.2188$$

Without replacement:

$p_1 = \frac{1}{8}$ and $q_1 = \frac{7}{8}$ on the first of the two draws. The batch number is now 39 for the second draw, thus $p_2 = \frac{5}{39}$ and $q_2 = \frac{35}{39}$

$$p_1q_2 + q_1p_2 = \frac{1}{8} \times \frac{35}{39} + \frac{7}{8} \times \frac{5}{39} = \frac{35+35}{312} = \frac{70}{312} \text{ or } 0.2244$$

Problem 8 A box contains seventy-four brass washers, eighty-six steel washers and forty aluminium washers. Three washers are drawn at random from the box without replacement. Determine the probability that all three are steel washers.

Assume, for clarity of explanation, that a washer is drawn at random, then a second, then a third, (although this assumption does not affect the results obtained). The total number of washers if 74+86+40, i.e., 200.

The probability of randomly selecting a steel washer on the first draw is 86/200. There are now 85 steel washers in a batch of 199. The probability of randomly selecting a steel washer on the second draw is 85/199. There are now 84 steel washers in a batch of 198. The probability of randomly selecting a steel washer on the third draw is 84/198. Hence the probability of selecting a steel washer on the first draw and the second draw and the third draw is

$$\frac{86}{200} \times \frac{85}{199} \times \frac{84}{198} = \frac{307\,020}{3\,940\,200} = \mathbf{\frac{5\,117}{65\,670}} \text{ or } \mathbf{0.078}$$

Problem 9 For the box of washers given in *Problem 8* above, determine the probability that there are no aluminium washers drawn, when three washers are drawn at random from the box without replacement.

The probability of not drawing an aluminium washer on the first draw is 1–(40/200), i.e. 160/200. There are now 199 washers in the batch of which 159 are not aluminium washers. Hence, the probability of not drawing an aluminium washer on the second draw is 159/199. Similarly, the probability of not drawing an aluminium washer on the third draw is 158/198. Hence the probability of not drawing an aluminium washer on the first and second and third draws is

$$\frac{160}{200} \times \frac{159}{199} \times \frac{158}{198} = \frac{100\,488}{197\,010} = \mathbf{\frac{50\,244}{98\,505}} \text{ or } \mathbf{0.510}$$

Problem 10 For the box of washers in *Problem 8* above, find the probability that there are two brass washers and either a steel or an aluminium washer when three are drawn at random, without replacement.

Two brass washers (A) and one steel washer (B) can be obtained in any of the following ways:

1st draw	*2nd draw*	*3rd draw*
A	A	B
A	B	A
B	A	A

Two brass washers and one aluminium washer (C) can also be obtained in any of the following ways:

1st draw	*2nd draw*	*3rd draw*
A	A	C
A	C	A
C	A	A

Thus there are six possible ways of achieving the combinations specified. If A represents a brass washer, B a steel washer and C an aluminium washer, then the combinations and their probabilities are as shown:

DRAW First	Second	Third	*PROBABILITY*
A	A	B	$\frac{74}{200} \times \frac{73}{199} \times \frac{86}{198} = 0.0590$
A	B	A	$\frac{74}{200} \times \frac{86}{199} \times \frac{73}{198} = 0.0590$
B	A	A	$\frac{86}{200} \times \frac{74}{199} \times \frac{73}{198} = 0.0590$
A	A	C	$\frac{74}{200} \times \frac{73}{199} \times \frac{40}{198} = 0.0274$
A	C	A	$\frac{74}{200} \times \frac{40}{199} \times \frac{73}{198} = 0.0274$
C	A	A	$\frac{40}{200} \times \frac{74}{199} \times \frac{73}{198} = 0.0274$

The probability of having the first combination **or** the second, **or** the third and so on is given by the sum of the probabilities, i.e., by $3 \times 0.0590+3 \times 0.0274$, that is, **0.2592**

C. FURTHER PROBLEMS ON PROBABILITY

1 In a batch of 45 lamps there are 10 faulty lamps. If one lamp is drawn at random, find the probability of it being (a) faulty and (b) satisfactory.

$$\left[(a)\ \frac{2}{9};\ (b)\ \frac{7}{9}\right]$$

2 A box of fuses are all of the same shape and size and comprises 23 2-A fuses, 47 5-A fuses and 69 13-A fuses. Determine the probability of selecting at random (a) a 2-A fuse, (b) a 5-A fuse and (c) a 13-A fuse.

$$\left[(a)\ \frac{23}{139} \text{ or } 0.1655;\ (b)\ \frac{47}{139} \text{ or } 0.3381;\ (c)\ \frac{69}{139} \text{ or } 0.4964\right]$$

3 (a) Find the probability of having a 2 upwards when throwing a fair 6-sided dice.
(b) Find the probability of having a 5 upwards when throwing a fair 6-sided dice.
(c) Determine the probability of having a 2 and then a 5 on two throws of a fair 6-sided dice.

$$\left[(a)\ \frac{1}{6};\ (b)\ \frac{1}{6};\ (c)\ \frac{1}{36}\right]$$

4 The probability of event A happening is 3/5 and the probability of event B happening is 2/3. Calculate the probabilities of (a) both A and B happening, (b) only event A happening, i.e. event A happening and event B not happening, (c) only event B happening and (d) either A, or B, or A and B happening.

$$\left[(a)\ \frac{2}{5};\ (b)\ \frac{1}{5};\ (c)\ \frac{4}{15};\ (d)\ \frac{13}{15}\right]$$

5 When testing 1000 soldered joints, 4 failed during a vibration test and 5 failed due to having a high resistance. Determine the probability of a joint failing due to (a) vibration, (b) high resistance, (c) vibration or high resistance and (d) vibration and high resistance.

$$\left[(a)\ \frac{1}{250};\ (b)\ \frac{1}{200};\ (c)\ \frac{9}{1000};\ (d)\ \frac{1}{50\,000}\right]$$

6 The probability that component A will operate satisfactorily for 5 years is 4/5 and that B will operate satisfactorily over the same period of time is 3/4. Find the probabilities that in a 5 year period:
(a) both components operate satisfactorily,
(b) only component A will operate satisfactorily, and
(c) only component B will operate satisfactorily.

$$\left[(a)\ \frac{3}{5};\ (b)\ \frac{1}{5};\ (c)\ \frac{3}{20}\right]$$

7 In a particular street, 80% of the houses have telephones. If two houses selected at random are visited, calculate the probabilities that (a) they both have a telephone and (b) one has a telephone but the other does not have a telephone.

$$\left[(a)\ \frac{16}{25};\ (b)\ \frac{8}{25}\right]$$

8 Veroboard pins are packed in packets of 20 by a machine. In a thousand packets, 40 have less than 20 pins. Find the probability that if 2 packets are chosen at random, one will contain less than 20 pins and the other will contain 20 pins or more.

$$\left[\frac{48}{625}\right]$$

9 A batch of 1-kW fire elements contain 16 which are within a power tolerance and 4 which are not. If 3 elements are selected at random from the batch, calculate the probabilities that (a) all three are within the power tolerance and (b) two are within but one is not within the power tolerance.

$$\left[\text{(a) } \frac{28}{57}; \text{ (b) } \frac{8}{19}\right]$$

10 The statistics on numbers of boys and girls of a group of families each having 3 children are analysed. Assuming equal probability for the birth of a boy or a girl, determine the percentage of the group likely to have (a) two boys and a girl, (b) at least one boy, (c) no girls and (d) at most two girls.

[(a) 12.5%; (b) 87.5%; (c) 12.5%; (d) 87.5%]

11 An amplifier is made up of three transistors, A, B and C. The probabilities of A, B or C being defective are 1/20, 1/25 and 1/50 respectively. Calculate the percentage of amplifiers produced (a) which work satisfactorily and (b) which have just one defective transistor. [(a) 89.38%; (b) 10.25%]

12 A box contains 14 40-W lamps, 28 60-W lamps and 58 25-W lamps, all the lamps being of the same shape and size. Three lamps are drawn at random from the box, first one, then a second, then a third. Determine the probabilities of: (a) getting one 25-W, one 40-W and one 60-W lamp, with replacement, (b) getting one 25-W, one 40-W and one 60-W lamp without replacement, (c) getting either one 25-W and two 40-W or one 60-W and two 40-W lamps with replacement and (d) getting either one 25-W and two 40-W or one 60-W and two 40-W lamps without replacement.

$$\left[\begin{array}{ll}\text{(a) } \dfrac{1421}{62\,500} \text{ or } 0.0227; & \text{(b) } \dfrac{2842}{121\,275} \text{ or } 0.0234; \\ \text{(c) } \dfrac{1421}{125\,000} \text{ or } 0.0114; & \text{(d) } \dfrac{11\,739}{242\,550} \text{ or } 0.0484\end{array}\right]$$

11 The binomial theorem

A. MAIN POINTS CONCERNED WITH THE BINOMIAL THEOREM

1 A **binomial expression** is one which contains two terms connected by a plus or minus sign. Thus $(p+q)$, $(a+x)^2$, $(2x+y)^3$ are examples of binomial expressions.

2 Expanding $(a+x)^n$ for integer values of n from 0 to 6 gives the following results:

$$(a+x)^0 = 1$$
$$(a+x)^1 = a+x$$
$$(a+x)^2 = (a+x)(a+x) = a^2+2ax+x^2$$
$$(a+x)^3 = (a+x)^2(a+x) = a^3+3a^2x+3ax^2+x^3$$
$$(a+x)^4 = (a+x)^3(a+x) = a^4+4a^3x+6a^2x^2+4ax^3+x^4$$
$$(a+x)^5 = (a+x)^4(a+x) = a^5+5a^4x+10a^3x^2+10a^2x^3+5ax^4+x^5$$
$$(a+x)^6 = (a+x)^5(a+x) = a^6+6a^5x+15a^4x^2+20a^3x^3+15a^2x^4+6ax^5+x^6$$

3 From the results of para. 2 the following patterns emerge:

(i) 'a' decreases in power moving from left to right.

(ii) 'x' increases in power moving from left to right.

(iii) The coefficients of each term of the expansions are symmetrical about the middle coefficient when n is even and symmetrical about the two middle coefficients when n is odd.

(iv) The coefficients are shown separately in *Table 1* and this arrangement is known as **Pascal's triangle**. A coefficient of a term may be obtained by adding the two adjacent coefficients immediately above in the previous row. This is shown by the triangles in *Table 1*, where, for example, $1+3 = 4$, $10+5 = 15$, and so on.

(v) Pascal's triangle method is used for expansion of the form $(a+x)^n$ for integer values of n less than about 8. The way in which Pascal's triangle can be used to obtain a binomial expansion is shown in *Problems 1 and 2.*

4 The **binomial theorem** is a formula for raising a binomial expression to any power without lengthy multiplication. The general binomial expansion of $(a+x)^n$ is given by:

$$(a+x)^n = a^n + na^{n-1}x + \frac{n(n-1)}{2!}a^{n-2}x^2 + \frac{n(n-1)(n-2)}{3!}a^{n-3}x^3 + \ldots\ldots + x^n$$

where 3! denotes $3 \times 2 \times 1$ and is termed 'factorial 3'.

With the binomial theorem n may be a fraction, a decimal fraction or a positive or negative integer. (See *Problems 3 and 4.*)

TABLE 1

$(a+x)^0$	1
$(a+x)^1$	1 1
$(a+x)^2$	1 2 1
$(a+x)^3$	1 3 3 1
$(a+x)^4$	1 4 6 4 1
$(a+x)^5$	1 5 10 10 5 1
$(a+x)^6$	1 6 15 20 15 6 1

5 The r'th term of the expansion $(a+x)^n$ is:

$$\frac{n(n-1)(n-2)\ldots\ldots \text{ to } (r-1) \text{ terms}}{(r-1)!}\, a^{n-(r-1)}x^{r-1}$$

(see *Problems 5 and 6*).

6 If $a = 1$ in the binomial expansion of $(a+x)^n$ then:

$$(1+x)^n = 1+nx+\frac{n(n-1)}{2!}x^2+\frac{n(n-1)(n-2)}{3!}x^3+\ldots\ldots\ldots,$$

which is valid for $-1<x<1$.

$(a + b)$ may be re-written as $a\left(1+\frac{b}{a}\right)$ and $(a+b)^n = \left[a\left(1+\frac{b}{a}\right)\right]^n$

$$= a^n\left(1+\frac{b}{a}\right)^n$$

When x is small compared with 1 then: $(1+x)^n \simeq 1+nx$.
(See *Problems 7 to 9.*)

7 Binomial expansions may be used for numerical approximations, for calculations with small variations and in probability theory (see chapter 24).
(See *Problems 10 and 11.*)

B WORKED PROBLEMS ON THE BINOMIAL THEOREM

Problem 1 Use the Pascal's triangle method to determine the expansion of $(a+x)^7$.

From *Table 1*, the row of Pascal's triangle corresponding to $(a+x)^6$ is as shown in (1) below. Adding adjacent coefficients gives the coefficients of $(a+x)^7$ as shown in (2) below.

	1 6	15		20 15		6	1	(1)
1	7	21	35	35	21	7	1	(2)

The first and last terms of the expansion of $(a+x)^7$ are a^7 and x^7 respectively. The powers of 'a' decrease and the powers of 'x' increase moving from left to right. Hence $\mathbf{(a+x)^7 = a^7+7a^6x+21a^5x^2+35a^4x^3+35a^3x^4+21a^2x^5+7ax^6+x^7}$.

Problem 2 Determine, using Pascal's triangle method, the expansion of $(2p-3q)^5$.

Comparing $(2p-3q)^5$ with $(a+x)^5$ shows that $a = 2p$ and $x = -3q$.
Using Pascal's triangle method:

$$(a+x)^5 = a^5 + 5a^4x + 10a^3x^2 + 10a^2x^3 \ldots$$

Hence $(2p-3q)^5 = (2p)^5 + 5(2p)^4(-3q) + 10(2p)^3(-3q)^2 + 10(2p)^2(-3q)^3 + 5(2p)(-3q)^4 + (-3q)^5$

i.e. $\mathbf{(2p-3q)^5 = 32p^5 - 240p^4q + 720p^3q^2 - 1080p^2q^3 + 810pq^4 - 243q^5}$

Problem 3 Use the binomial theorem to determine the expansion of $(2+x)^7$.

The binomial expansion of $(a+x)^n = a^n + na^{n-1}x + \dfrac{n(n-1)}{2!}a^{n-2}x^2 + \dfrac{n(n-1)(n-2)}{3!}a^{n-3}x^3 + \ldots$

When $a = 2$ and $n = 7$:

$$(2+x)^7 = 2^7 + 7(2)^6x + \frac{(7)(6)}{(2)(1)}(2)^5x^2 + \frac{(7)(6)(5)}{(3)(2)(1)}(2)^4x^3 + \frac{(7)(6)(5)(4)}{(4)(3)(2)(1)}(2)^3x^4 + \frac{(7)(6)(5)(4)(3)}{(5)(4)(3)(2)(1)}(2)^2x^5 + \frac{(7)(6)(5)(4)(3)(2)}{(6)(5)(4)(3)(2)(1)}(2)x^6 + \frac{(7)(6)(5)(4)(3)(2)(1)}{(7)(6)(5)(4)(3)(2)(1)}x^7$$

i.e. $\mathbf{(2+x)^7 = 128 + 448x + 672x^2 + 560x^3 + 280x^4 + 84x^5 + 14x^6 + x^7}$.

Problem 4 Expand $\left(c - \dfrac{1}{c}\right)^5$ using the binomial theorem.

Using the binomial expansion of $(a+x)^n$, where $a = c$, $x = -\dfrac{1}{c}$ and $n = 5$ gives:

$$\left(c - \frac{1}{c}\right)^5 = c^5 + 5c^4\left(-\frac{1}{c}\right) + \frac{(5)(4)}{(2)(1)}c^3\left(-\frac{1}{c}\right)^2 + \frac{(5)(4)(3)}{(3)(2)(1)}c^2\left(-\frac{1}{c}\right)^3 + \frac{(5)(4)(3)(2)}{(4)(3)(2)(1)}c\left(-\frac{1}{c}\right)^4 + \frac{(5)(4)(3)(2)(1)}{(5)(4)(3)(2)(1)}\left(-\frac{1}{c}\right)^5$$

i.e. $\mathbf{\left(c - \dfrac{1}{c}\right)^5 = c^5 - 5c^3 + 10c - \dfrac{10}{c} + \dfrac{5}{c^3} - \dfrac{1}{c^5}}$

Problem 5 Without fully expanding $(3+x)^7$, determine the fifth term.

The r'th term of the expansion $(a+x)^n$ is given by:

$$\frac{n(n-1)(n-2) \ldots\ldots \text{ to } (r-1) \text{ terms}}{(r-1)!}a^{n-(r-1)}x^{r-1}.$$

Substituting $n = 7$, $a = 3$ and $r-1 = 5-1 = 4$ gives: $\dfrac{(7)(6)(5)(4)}{(4)(3)(2)(1)}(3)^{7-4}x^4$

i.e. the fifth term of $(3+x)^7 = 35(3)^3x^4 = \mathbf{945x^4}$

Problem 6 Find the middle term of $\left(2p - \dfrac{1}{2q}\right)^{10}$.

In the expansion of $(a+x)^{10}$ there are 10+1, i.e. 11 terms. Hence the middle term is the sixth. Using the general expression for the r'th term where $a = 2p$, $x = -\dfrac{1}{2q}$, $n = 10$ and $r-1 = 5$ gives:

$$\frac{(10)(9)(8)(7)(6)}{(5)(4)(3)(2)(1)}(2p)^{10-5}\left(-\frac{1}{2q}\right)^5 = 252(32p^5)\left(-\frac{1}{32q^5}\right)$$

Hence the middle term of $\left(2p - \frac{1}{2q}\right)^{10}$ is $\mathbf{-252\frac{p^5}{q^5}}$

Problem 7 Evaluate $(1.002)^9$ using the binomial theorem correct to (a) 3 decimal places and (b) 7 significant figures.

$$(1+x)^n = 1+nx+\frac{n(n-1)}{2!}x^2+\frac{n(n-1)(n-2)}{3!}x^3+\ldots\ldots\ldots\ldots$$

$(1.002)^9 = (1+0.002)^9$. Substituting $x = 0.002$ and $n = 9$ in the general expansion for $(1+x)^n$ gives:

$$\begin{aligned}(1+0.002)^9 &= 1+9(0.002)+\frac{(9)(8)}{(2)(1)}(0.002)^2+\frac{(9)(8)(7)}{(3)(2)(1)}(0.002)^3+\ldots\\ &= 1+0.018+0.000144+0.000000672+\ldots\ldots\\ &= 1.018144672\ldots\ldots\end{aligned}$$

Hence $(1.002)^9$ = **1.018, correct to 3 decimal places**
= **1.018145, correct to 7 significant figures.**

Problem 8 Evaluate $(0.97)^6$ correct to 4 significant figures using the binomial expansion.

$(0.97)^6$ is written as $(1-0.03)^6$. Using the expansion of $(1+x)^n$ where $n = 6$ and $x = -0.03$ gives:

$$\begin{aligned}(1-0.03)^6 &= 1+6(-0.03)+\frac{(6)(5)}{(2)(1)}(-0.03)^2+\frac{(6)(5)(4)}{(3)(2)(1)}(-0.03)^3\\ &\quad+\frac{(6)(5)(4)(3)}{(4)(3)(2)(1)}(-0.03)^4+\ldots\\ &= 1-0.18+0.0135-0.00054+0.00001215-\ldots\ldots\ldots\\ &\simeq 0.83297215\end{aligned}$$

i.e. $\mathbf{(0.97)^6}$ = **0.8330, correct to 4 significant figures.**

Problem 9 Determine the value of $(3.039)^4$, correct to 6 significant figures using the binomial theorem.

$(3.039)^4$ may be written in the form $(1+x)^n$ as:

$$(3.039)^4 = (3+0.039)^4 = \left[3\left(1+\frac{0.039}{3}\right)\right]^4 = 3^4(1+0.013)^4$$

$$\begin{aligned}(1+0.013)^4 &= 1+4(0.013)+\frac{(4)(3)}{(2)(1)}(0.013)^2+\frac{(4)(3)(2)}{(3)(2)(1)}(0.013)^3+\ldots\ldots\\ &= 1+0.052+0.001014+0.000008788+\ldots\ldots\ldots\\ &= 1.0530228 \text{ correct to 8 significant figures.}\end{aligned}$$

Hence $(3.039)^4 = 3^4(1.0530228)$ = **85.2948, correct to 6 significant figures.**

Problem 10 The radius of a cylinder is reduced by 4% and its height is increased by 2%. Determine the approximate percentage change in (a) its volume and (b) its curved surface area, (neglecting the products of small quantities).

Volume of cylinder = $\pi r^2 h$

Let r and h be the original values of radius and height.

The new values are $0.96r$ or $(1-0.04)r$ and $1.02h$ or $(1+0.02)h$

(a) New volume $= \pi[(1-0.04)r]^2[(1+0.02)h]$

$= \pi r^2 h(1-0.04)^2(1+0.02)$

Now $(1-0.04)^2 = 1-(2 \times 0.04)+(0.04)^2 = (1-0.08)$, neglecting powers of small terms.

Hence new volume $\simeq \pi r^2 h(1-0.08)(1+0.02)$

$\simeq \pi r^2 h(1-0.08+0.02)$, neglecting products of small terms

$\simeq \pi r^2 h(1-0.06)$ or $0.94\pi r^2 h$, i.e., 94% of the original volume.

Hence the volume is reduced by approximately 6%.

(b) Curved surface area of cylinder $= 2\pi rh$

New surface area $= 2\pi[(1-0.04)r][(1+0.02)h]$

$= 2\pi rh(1-0.04)(1+0.02)$

$\simeq 2\pi rh(1-0.04+0.02)$, neglecting products of small terms

$\simeq 2\pi rh(1-0.02)$ or $0.98(2\pi rh)$, i.e., 98% of the original surface area.

Hence the curved surface area is reduced by approximately 2%.

Problem 11 The second moment of area of a rectangle through its centroid is given by $bl^3/12$. Determine the approximate change in the second moment of area if b is increased by 3.5% and l is reduced by 2.5%.

New values of b and l are $(1+0.035)b$ and $(1-0.025)l$ respectively.

New second moment of area $= \frac{1}{12}[(1+0.035)b][(1-0.025)l]^3$

$= \frac{bl^3}{12}(1+0.035)(1-0.025)^3$

$\simeq \frac{bl}{12}(1+0.035)(1-0.075)$, neglecting powers of small terms

$\simeq \frac{bl^3}{12}(1+0.035-0.075)$, neglecting products of small terms

$\simeq \frac{bl^3}{12}(1-0.040)$ or $(0.96)\frac{bl^3}{12}$, i.e. 96% of the original second moment of area

Hence the second moment of area is reduced by approximately 4%.

C. FURTHER PROBLEMS ON THE BINOMIAL THEOREM

1 Use Pascal's triangle to expand $(x-y)^7$.

$[x^7-7x^6y+21x^5y^2-35x^4y^3+35x^3y^4-21x^2y^5+7xy^6-y^7]$

2 Expand $(2a+3b)^5$ using Pascal's triangle.

$[32a^5+240a^4b+720a^3b^2+1080a^2b^3+810ab^4+243b^5]$

3 Use the binomial theorem to expand (a) $(a+2x)^4$; (b) $(2-x)^6$.

$\begin{bmatrix}\text{(a) } a^4+8a^3x+24a^2x^2+32ax^3+16x^4; \\ \text{(b) } 64-192x+240x^2-160x^3+60x^4-12x^5+x^6\end{bmatrix}$

4 Expand (a) $(2x-3y)^4$; (b) $\left(2x+\frac{2}{x}\right)^5$.

$$\left[\begin{array}{l}(a)\ 16x^4-96x^3y+216x^2y^2-216xy^3+81y^4;\\(b)\ 32x^5+160x^3+320x+\frac{320}{x}+\frac{160}{x^3}+\frac{32}{x^5}\end{array}\right]$$

5 Expand $(p+2q)^{11}$ as far as the fifth term.

[$p^{11}+22p^{10}q+220p^9q^2+1320p^8q^3+5280p^7q^4$]

6 Determine the sixth term of $(3p+\frac{q}{3})^{13}$ [$34749p^8q^5$]

7 Determine the middle term of $(2a-5b)^8$. [$700\,000a^4b^4$]

8 Use the binomial theorem to determine, correct to 4 decimal places:
(a) $(1.003)^8$; (b) $(1.042)^7$ [(a) 1.0243; (b) 1.3337]

9 Use the binomial theorem to determine, correct to 5 significant figures:
(a) $(0.98)^7$; (b) $(2.01)^9$ [(a) 0.86813; (b) 535.51]

10 Evaluate $(4.044)^6$ correct to 3 decimal places. [4373.880]

11 Pressure p and volume v are related by $pv^3 = c$, where c is a constant. Determine the approximate percentage change in c when p is increased by 3% and v decreased by 1.2%. [0.6% decrease]

12 Kinetic energy is given by $\frac{1}{2}mv^2$. Determine the approximate change in the kinetic energy when mass m is increased by 2.5% and the velocity v is reduced by 3%. [3.5% decrease]

13 An error of +1.5% was made when measuring the radius of a sphere. Ignoring the products of small quantities determine the approximate error in calculating (a) the volume, and (b) the surface area. [(a) 4.5% increase; (b) 3.0% increase]

14 The power developed by an engine is given by $I = k\,PLAN$, where k is a constant. Determine the approximate percentage change in the power when P and A are each increased by 2.5% and L and N are each decreased by 1.4%.

[2.2% increase]

15 The radius of a cone is increased by 2.7% and its height reduced by 0.9%. Determine the approximate percentage change in its volume, neglecting the products of small terms. [4.5% increase]

12 Arithmetical and geometric progressions

A. MAIN POINTS CONCERNED WITH ARITHMETICAL AND GEOMETRIC PROGRESSIONS

Arithmetical progressions

1 When a sequence has a constant difference between successive terms it is called an **arithmetical progression** (often abbreviated to AP).
Examples include (i) 1, 4, 7, 10, 13, ... where the **common difference** is 3
and (ii) $a, a + d, a + 2d, a + 3d$, ... where the common difference is d.

2 If the first term of an AP is 'a' and the common difference is 'd' then the n'th term is $a + (n - 1)d$. In example (i) of para. 1, the 7'th term is given by $1 + (7 - 1)3 = 19$, which may be readily checked.

3 The sum S of an AP can be obtained by multiplying the average of all the terms by the number of terms.

The average of all the terms $= \dfrac{a + l}{2}$,

where 'a' is the first term and l is the last term,
i.e. $l = a + (n - 1)d$, for n terms.

Hence the sum of n terms, $S_n = n\left(\dfrac{a + l}{2}\right) = \dfrac{n}{2}\left\{a + [a + (n - 1)d]\right\}$

$$\text{i.e.}\quad S_n = \frac{n}{2}[2a + (n - 1)d]$$

For example, the sum of the first 7 terms of the series 1, 4, 7, 10, 13, ... is

given by $S_7 = \dfrac{7}{2}[2(1) + (7 - 1)3]$, since $a = 1$ and $d = 3$,

$$= \frac{7}{2}[2 + 18] = \frac{7}{2}[20] = \mathbf{70}$$

(See *Problems 1 to 9*)

Geometric progressions

4 When a sequence has a constant ratio between successive terms it is called a **geometric progression** (often abbreviated to GP). The constant is called the **common ratio**, r.
Examples include (i) 1, 2, 4, 8, ... where the common ratio is 2
and (ii) a, ar, ar^2, ar^3, ... where the common ratio is r.

5 If the first term of a GP is 'a' and the common ratio is r, then the n'th term is ar^{n-1}, which can be readily checked from the examples in para. 4.
For example, the 8'th term of the GP 1, 2, 4, 8, ... is $(1)(2)^7 = \mathbf{128}$, since $a = 1$ and $r = 2$.

6 Let a GP be $a, ar, ar^2, ar^3, \ldots . ar^{n-1}$,
then the sum of n terms, $S_n = a + ar + ar^2 + ar^3 + \ldots . + ar^{n-1}$ . . . (1)
Multiplying throughout by r gives: $rS_n = ar + ar^2 + ar^3 + ar^4 + \ldots \ldots ar^{n-1} + ar^n$ . . . . . .(2)
Subtracting equation (2) from equation (1) gives $S_n - rS_n = a - ar^n$
i.e. $S_n(1 - r) = a(1 - r^n)$

Thus the sum of n terms, $S_n = \dfrac{a(1 - r^n)}{(1 - r)}$, which is valid when $r < 1$.

Subtracting equation (1) from equation (2) gives $S_n = \dfrac{a(r^n - 1)}{(r - 1)}$, which is valid when $r > 1$.

For example, the sum of the first 8 terms of the G.P. 1, 2, 4, 8, 16, ... is given by $S_8 = \dfrac{1(2^8 - 1)}{(2 - 1)}$, since $a = 1$ and $r = 2$,

i.e. $S_8 = \dfrac{1(256 - 1)}{1} = \mathbf{255}$.

7 When the common ratio r of a GP is less than unity, the sum of n terms,
$S_n = \dfrac{a(1 - r^n)}{(1 - r)}$, which may be written as $S_n = \dfrac{a}{(1 - r)} - \dfrac{ar^n}{(1 - r)}$. Since $r < 1$,
r^n becomes less as n increases, i.e., $r^n \to 0$ as $n \to \infty$. Hence $\dfrac{ar^n}{(1 - r)} \to 0$ as $n \to \infty$.

Thus $S_n \to \dfrac{a}{(1 - r)}$ as $n \to \infty$. The quantity $\dfrac{a}{(1 - r)}$ is called the **sum to infinity**,

S_∞, and is the limiting value of the sum of an infinite number of terms,

i.e., $S_\infty = \dfrac{a}{1 - r}$, which is valid when $-1 < r < 1$.

For example, the sum to infinity of the G.P. $1 + \dfrac{1}{2} + \dfrac{1}{4} + \ldots\ldots$ is $S_\infty = \dfrac{1}{1 - 1½}$,

since $a = 1$ and $r = \dfrac{1}{2}$, i.e., $S_\infty = 2$

(See *Problems 10 to 17*)

B. WORKED PROBLEMS ON ARITHMETICAL AND GEOMETRIC PROGRESSIONS

Problem 1 Determine (a) the ninth, and (b) the sixteenth term of the series 2, 7, 12, 17, ...

2, 7, 12, 17, ... is an arithmetic progression with a common difference, d, of 5.
(a) From para. 2, the n'th term of an A.P. is given by $a + (n - 1)d$.
Since the first term $a = 2$, $d = 5$ and $n = 9$,
then the 9'th term is $2 + (9 - 1)5 = 2 + (8)(5) = 2 + 40 = \mathbf{42}$
(b) The 16'th term is $2 + (16 - 1)5 = 2 + (15)(5) = 2 + 75 = \mathbf{77}$

Problem 2 The 6'th term of an A.P. is 17 and the 13'th term is 38. Determine the 19'th term.

From para. 2, the n'th term of an A.P. is $a + (n - 1)d$.
The 6'th term is $a + 5d = 17$ (1)
The 13'th term is $a + 12d = 38$ (2)

Equation (2) – equation (1) gives $7d = 21$, from which, $d = \frac{21}{7} = 3$

Substituting in equation (1) gives: $a + 15 = 17$, from which, $a = 2$.
Hence the 19'th term is $a + (n - 1)d = 2 + (19 - 1)3 = 2 + (18)(3) = 2 + 54 =$ **56**

Problem 3 Determine the number of the term whose value is 22 in the series 2½, 4, 5½, 7, ...

2½, 4, 5½, 7, ... is an A.P. where $a = 2½$ and $d = 1½$
Hence if the n'th term is 22 then: $a + (n - 1)d = 22$
i.e. $2½ + (n - 1)1½ = 22$; $(n - 1)1½ = 22 - 2½ = 19½$

$$n - 1 = \frac{19½}{1½} = 13 \text{ and } n = 13 + 1 = 14$$

i.e. **the 14'th term of the A.P. is 22**

Problem 4 Find the sum of the first 12 terms of the series 5, 9, 13, 17, ...

5, 9, 13, 17, ... is an A.P. where $a = 5$ and $d = 4$

From para. 3, the sum of n terms of an A.P., $S_n = \frac{n}{2}[2a + (n - 1)d]$

Hence the sum of the first 12 terms, $S_{12} = \frac{12}{2}[2(5) + (12 - 1)4]$

$= 6[10 + 44] = 6(54) =$ **324**

Problem 5 Find the sum of the first 21 terms of the series 3.5, 4.1, 4.7, 5.3, ...

3.5, 4.1, 4.7, 5.3, ... is an A.P. where $a = 3.5$ and $d = 0.6$

The sum of the first 21 terms, $S_{21} = \frac{n}{2}[2a + (n - 1)d]$

$$= \frac{21}{2}[2(3.5) + (21 - 1)0.6] = \frac{21}{2}[7 + 12]$$

$$= \frac{21}{2}(19) = \frac{399}{2} = \mathbf{199.5}$$

Problem 6 The sum of 7 terms of an A.P. is 35 and the common difference is 1.2. Determine the first term of the series.

$n = 7$, $d = 1.2$ and $S_7 = 35$.

Since the sum of n terms of an A.P. is given by $S_n = \frac{n}{2}[2a + (n - 1)d]$, then

$$35 = \frac{7}{2}[2a + (7 - 1)1.2] = \frac{7}{2}[2a + 7.2]$$

Hence $\frac{35 \times 2}{7} = 2a + 7.2$

$$10 = 2a + 7.2$$

Thus $2a = 10 - 7.2 = 2.8$, from which $a = \frac{2.8}{2} = 1.4$

i.e. **the first term, $a = 1.4$**

Problem 7 Three numbers are in arithmetic progression. Their sum is 15 and their product is 80. Determine the three numbers.

Let the three numbers be $(a - d)$, a and $(a + d)$.
Then $(a - d) + a + (a + d) = 15$, i.e., $3a = 15$, from which, $a = 5$
Also, $a(a - d)(a + d) = 80$, i.e., $a(a^2 - d^2) = 80$
Since $a = 5$, $5(5^2 - d^2) = 80$

$$125 - 5d^2 = 80$$
$$125 - 80 = 5d^2$$
$$45 = 5d^2$$

from which, $d^2 = \frac{45}{5} = 9$. Hence $d = \sqrt{9} = \pm 3$

The three numbers are thus $(5 - 3)$, 5 and $(5 + 3)$ **i.e. 2, 5 and 8**

Problem 8 Find the sum of all the numbers between 0 and 207 which are exactly divisible by 3

The series 3, 6, 9, 12, ... 207 is an A.P. whose first term $a = 3$ and common difference $d = 3$
The last term is $a + (n - 1)d = 207$

i.e. $3 + (n - 1)3 = 207$, from which $(n - 1) = \frac{207 - 3}{3} = 68$

Hence $n = 68 + 1 = 69$

The sum of all 69 terms is given by $S_{69} = \frac{n}{2}[2a + (n - 1)d]$

$$= \frac{69}{2}[2(3) + (69 - 1)3]$$

$$= \frac{69}{2}[6 + 204] = \frac{69}{2}(210) = \mathbf{7245}$$

Problem 9 The first, twelth and last term of an arithmetic progression are 4, 31½ and 376½ respectively. Determine (a) the number of terms in the series, (b) the sum of all the terms and (c) the 80'th term

(a) Let the A.P. be $a, a + d, a + 2d, \ldots\ldots, a + (n - 1)d$, where $a = 4$
The 12'th term is $a + (12 - 1)d = 31\frac{1}{2}$

i.e. $4 + 11d = 31\frac{1}{2}$, from which, $11d = 31\frac{1}{2} - 4 = 27\frac{1}{2}$

Hence $d = \frac{27\frac{1}{2}}{11} = 2\frac{1}{2}$

The last term is $a + (n - 1)d$

i.e. $4 + (n - 1)2\frac{1}{2} = 376\frac{1}{2}$

$$(n - 1) = \frac{376\frac{1}{2} - 4}{2\frac{1}{2}} = \frac{372\frac{1}{2}}{2\frac{1}{2}} = 149$$

Hence the number of terms in the series, $n = 149 + 1 = 150$

(b) Sum of all the terms, $S_{150} = \frac{n}{2}[2a + (n-1)d] = \frac{150}{2}[2(4) + (150-1)2\frac{1}{2}]$

$= 75[8 + (149)(2\frac{1}{2})] = 75[8 + 372.5] = 75(380.5)$

$= \mathbf{28\,537\frac{1}{2}}$

(c) The 80'th term is $a + (n-1)d = 4 + (80-1)2\frac{1}{2}$

$= 4 + (79)(2\frac{1}{2}) = 4 + 197.5 = \mathbf{201\frac{1}{2}}$

Problem 10 Determine the tenth term of the series 3, 6, 12, 24, ...

3, 6, 12, 24, ... is a geometric progression with a common ratio r of 2. From para. 5, the n'th term of a G.P. is ar^{n-1}, where a is the first term. Hence the 10'th term is $(3)(2)^{10-1} = (3)(2)^9 = 3(512) =$ **1536**

Problem 11 Find the sum of the first 7 terms of the series, ½, 1½, 4½, 13½, ...

½, 1½, 4½, 13½, ... is a G.P. with common ratio $r = 3$

From para. 6, the sum of n terms, $S_n = \frac{a(r^n - 1)}{(r-1)}$

Hence $S_7 = \frac{\frac{1}{2}(3^7 - 1)}{(3-1)} = \frac{\frac{1}{2}(2187 - 1)}{2} = \mathbf{546\frac{1}{2}}$

Problem 12 The first term of a geometric progression is 12 and the fifth term is 55. Determine the 8'th term and the 11'th term

The 5'th term is given by $ar^4 = 55$, where the first term $a = 12$

Hence $r^4 = \frac{55}{a} = \frac{55}{12}$ and $r = \sqrt[4]{\left(\frac{55}{12}\right)} = 1.4632$

The 8'th term is $ar^7 = (12)(1.4632)^7 =$ **172.3**

The ll'th term is $ar^{10} = (12)(1.4632)^{10} =$ **539.8**

Problem 13 Which term of the series 2187, 729, 243, ... is $\frac{1}{9}$?

2187, 729, 243 . . . . . . is a G.P. with a common ratio $r = \frac{1}{3}$ and first term $a = 2187$.

From para. 5, the n'th term of a G.P. is given by ar^{n-1}.

Hence $\frac{1}{9} = (2187)(\frac{1}{3})^{n-1}$ from which $(\frac{1}{3})^{n-1} = \frac{1}{(9)(2187)} = \frac{1}{3^2 3^7} = \frac{1}{3^9} = \left(\frac{1}{3}\right)^9$

Thus $(n-1) = 9$, from which, $n = 9 + 1 = 10$

i.e. $\frac{1}{9}$ **is the 10'th term of the G.P.**

Problem 14 Find the sum of the first 9 terms of the series 72.0, 57.6, 46.08, ...

The common ratio, $r = \frac{ar}{a} = \frac{57.6}{72.0} = 0.8$ (also $\frac{ar^2}{ar} = \frac{46.08}{57.6} = 0.8$) hence the series is a GP.

From para. 6, the sum of 9 terms, $S_9 = \frac{a(1 - r^n)}{(1-r)} = \frac{72.0\,(1 - 0.8^9)}{(1 - 0.8)}$

$= \frac{72.0\,(1 - 0.1342)}{0.2} = \mathbf{311.7}$

Problem 15 Find the sum to infinity of the series $3, 1, \frac{1}{3}, \ldots\ldots$

$3, 1, \frac{1}{3}, \ldots\ldots$ is a G.P. of common ratio, $r = \frac{1}{3}$.

From para. 7, the sum to infinity, $S_\infty = \frac{a}{1-r} = \frac{3}{1-\frac{1}{3}} = \frac{3}{\frac{2}{3}} = \frac{9}{2} = 4\frac{1}{2}$

Problem 16 In a geometric progression the sixth term is 8 times the third term and the sum of the seventh and eighth terms is 192. Determine (a) the common ratio, (b) the first term, and (c) the sum of the fifth to eleventh terms, inclusive.

(a) Let the G.P. be $a, ar, ar^2, ar^3, \ldots\ldots, ar^{n-1}$
The 3rd term $= ar^2$ and the sixth term $= ar^5$.
The 6th term is 8 times the 3rd.
Hence $ar^5 = 8\,ar^2$ from which, $r^3 = 8, r = \sqrt[3]{8}$,
i.e. **the common ratio $r = 2$**

(b) The sum of the 7'th and 8'th terms is 192. Hence $ar^6 + ar^7 = 192$
Since $r = 2$, then $64a + 128a = 192$
$192a = 192$, from which a, **the first term, = 1**

(c) The sum of the 5'th to 11'th terms (inclusive) is given by:

$$S_{11} - S_4 = \frac{a(r^{11}-1)}{(r-1)} - \frac{a(r^4-1)}{(r-1)} = \frac{1(2^{11}-1)}{(2-1)} - \frac{1(2^4-1)}{(2-1)}$$

$$= (2^{11}-1) - (2^4-1) = 2^{11} - 2^4 = 2048 - 16 = \mathbf{2032}$$

Problem 17 A hire tool firm finds that their net return from hiring tools is decreasing by 10% per annum. If their net gain on a certain tool this year is £400, find the possible total of all future profits from this tool (assuming the tool lasts for ever).

The net gain forms a series: £400 + £400 × 0.9 + £400 × 0.9^2 + ..., which is a G.P. with $a = 400$ and $r = 0.9$

The sum to infinity, $S_\infty = \frac{a}{(1-r)} = \frac{400}{(1-0.9)}$ = **£4000 = total future profits**

C. FURTHER PROBLEMS ON ARITHMETICAL AND GEOMETRIC PROGRESSIONS

ARITHMETICAL PROGRESSIONS

1 Find the 11th term of the series 8, 14, 20, 26, ...

[68]

2 Find the 17th term of the series 11, 10.7, 10.4, 10.1, ...

[6.2]

3 The seventh term of a series is 29 and the eleventh term is 54. Determine the sixteenth term.

[85.25]

4 Find the 15th term of an arithmetic progression of which the first term is $2\frac{1}{2}$ and the tenth term is 16.

[$23\frac{1}{2}$]

5 Determine the number of the term which is 29 in the series 7, 9.2, 11.4, 13.6, ...

[11]

6 Find the sum of the first 11 terms of the series 4, 7, 10, 13, ...

[209]

7 Determine the sum of the series 6.5 + 8.0 + 9.5 + 11.0 + ... + 32.

[346.5]

8 The sum of 15 terms of an arithmetic progression is 202.5 and the common difference is 2. Find the first term of the series.

[$-\frac{1}{2}$]

9 Three numbers are in arithmetic progression. Their sum is 9 and their product is $20\frac{1}{4}$. Determine the three numbers.

[$1\frac{1}{2}$, 3, $4\frac{1}{2}$]

10 Find the sum of all the numbers between 5 and 250 which are exactly divisible by 4.

[7808]

11 Find the number of terms of the series 5, 8, 11, ... of which the sum is 1025.

[25]

12 Insert four terms between 5 and $22\frac{1}{2}$ to form an arithmetic progression.

[$8\frac{1}{2}$, 12, $15\frac{1}{2}$, 19]

13 The first, tenth and last terms of an arithmetic progression are 9, 40.5 and 425.5 respectively. Find (a) the number of terms; (b) the sum of all the terms and (c) the 70'th term.

[(a) 120; (b) 26070; (c) 250.5]

14 On commencing employment a man is paid a salary of £7200 per annum and receives annual increments of £350. Determine his salary in the 9'th year and calculate the total he will have received in the first 12 years.

[£10 000; £109 500]

15 An oil company bores a hole 80 m deep. Estimate the cost of boring if the cost is £30 for drilling the first metre with an increase in cost of £2 per metre for each succeeding metre.

[£8720]

GEOMETRIC PROGRESSIONS

16 Find the 10'th term of the series 5, 10, 20, 40, ...

[2560]

17 Determine the sum of the first 7 terms of the series $\frac{1}{4}$, $\frac{3}{4}$, $2\frac{1}{4}$, $6\frac{3}{4}$, ...

[273.25]

18 The first term of a geometric progression is 4 and the 6'th term is 128. Determine the 8'th and 11'th terms.

[512; 4096]

19 Which term of the series 3, 9, 27, ... is 59 049?

[10'th]

20 Find the sum of the first 7 terms of the series 2, 5, $12\frac{1}{2}$, ... (correct to 4 significant figures).

[812.5]

21 Determine the sum to infinity of the series 4, 2, 1, ...

[8]

22 Find the sum to infinity of the series $2\frac{1}{2}, -1\frac{1}{4}, \frac{5}{8}, \ldots\ldots$

[$1\frac{2}{3}$]

23 In a geometric progression the 5'th term is 9 times the 3rd term and the sum of the 6'th and 7'th terms is 1944. Determine (a) the common ratio, (b) the first term and (c) the sum of the 4'th to 10'th terms inclusive.

[(a) 3, (b) 2, (c) 59 022]

24 The value of a lathe originally valued at £3000 depreciates 15% per annum. Calculate its value after 5 years. The machine is sold when its value is less than £550. After how many years is the lathe sold?

[£1566; 11 years]

25 If the population of Great Britain is 55 million and is decreasing at 2.4% per annum, what will be the population in 6 years time?

[48.71 M]

26 100 grammes of a radioactive substance disintegrates at a rate of 3% per annum. How much of the substance is left after 12 years?

[71.53 g]

13 Reduction of non-linear laws to linear form

A. MAIN POINTS CONCERNED WITH THE REDUCTION OF NON-LINEAR LAWS TO LINEAR FORM

1 If the equation of a graph is of the form $y = mx + c$, where m and c are constants, the graph will always be a straight line, m representing the gradient and c the y-axis intercept. Thus, $y = 5x + 2$ represents a straight line of gradient 5 and y-axis intercept 2. Similarly, $y = -3x - 4$ represents a straight line of gradient -3 and y-axis intercept -4.

2 Frequently, the relationship between two variables, say x and y, is not a linear one, i.e. when x is plotted against y a curve results. In such cases the non-linear equation may be modified to the linear form, $y = mx + c$, so that the constants, and thus the law relating the variables can be determined. This technique is called **'determination of law'.**

3 Some examples of the reduction of equations to linear form include:

(i) $y = ax^2 + b$ compares with $Y = mX + c$, where $m = a$, $c = b$ and $X = x^2$. Hence y is plotted vertically against x^2 horizontally to produce a straight line graph of gradient 'a' and y-axis intercept 'b'.

(ii) $y = \frac{a}{x} + b$.

y is plotted vertically against $\frac{1}{x}$ horizontally to produce a straight line graph of gradient 'a' and y-axis intercept 'b'

(iii) $y = ax^2 + bx$.

Dividing both sides by x gives $\frac{y}{x} = ax + b$.

Comparing with $Y = mX + c$ shows that $\frac{y}{x}$ is plotted vertically against x horizontally to produce a straight line graph of gradient 'a' and $\frac{y}{x}$-axis intercept 'b'.

B. WORKED PROBLEMS ON REDUCING NON-LINEAR LAWS TO LINEAR FORM

Problem 1 Experimental values of x and y, shown below, are believed to be related by the law $y = ax^2 + b$. By plotting a suitable graph verify this law and determine approximate values of a and b.

x	1	2	3	4	5
y	9.8	15.2	24.2	36.5	53.0

If y is plotted against x a curve results and it is not possible to determine the values of constants a and b from the curve. Comparing $y = ax^2 + b$ with $Y = m X + c$ shows that y is to be plotted vertically against x^2 horizontally. A table of values is drawn up as shown below.

x	1	2	3	4	5
x^2	1	4	9	16	25
y	9.8	15.2	24.2	36.5	53.0

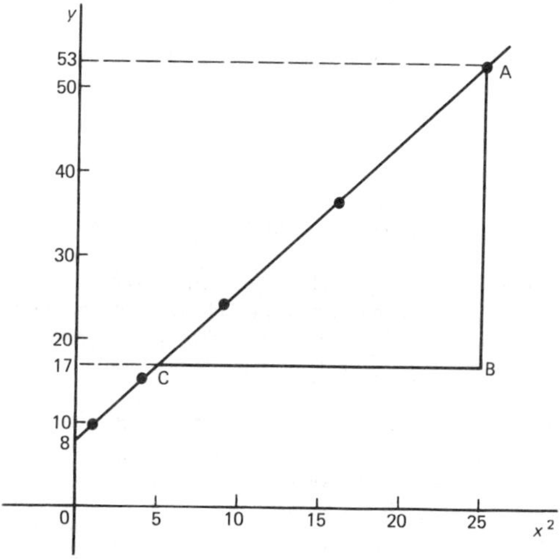

Fig 1

A graph of y against x^2 is shown in *Fig 1*, with the best straight line drawn through the points. Since a straight line graph results, the law is verified.

From the graph, gradient, $a = \dfrac{AB}{BC} = \dfrac{53 - 17}{25 - 5} = \dfrac{36}{20} = \mathbf{1.8}$,

and the y-axis intercept, $b = \mathbf{8.0}$

Hence the law of the graph is $y = \mathbf{1.8}\,x^2 + \mathbf{8.0}$

Problem 2 Values of load L newtons and distance d metres obtained experimentally are shown in the following table.

Load, L N	32.3	29.6	27.0	23.2	18.3	12.8	10.0	6.4
distance, d m	0.75	0.37	0.24	0.17	0.12	0.09	0.08	0.07

Verify that load and distance are related by a law of the form $L = \dfrac{a}{d} + b$ and determine approximate values of a and b. Hence calculate the load when the distance is 0.20 m and the distance when the load is 20 N.

Comparing $L = \frac{a}{d} + b$, i.e., $L = a\left(\frac{1}{d}\right) + b$, with $Y = mX + c$ shows that L is to be plotted vertically against $\frac{1}{d}$ horizontally. Another table of values is drawn up as shown below.

L	32.3	29.6	27.0	23.2	18.3	12.8	10.0	6.4
d	0.75	0.37	0.24	0.17	0.12	0.09	0.08	0.07
$\frac{1}{d}$	1.33	2.70	4.17	5.88	8.33	11.1	12.5	14.3

A graph of L against $\frac{1}{d}$ is shown in *Fig 2*. A straight line can be drawn through the points, which verifies that load and distance are related by a law of the form $L = \frac{a}{d} + b$.

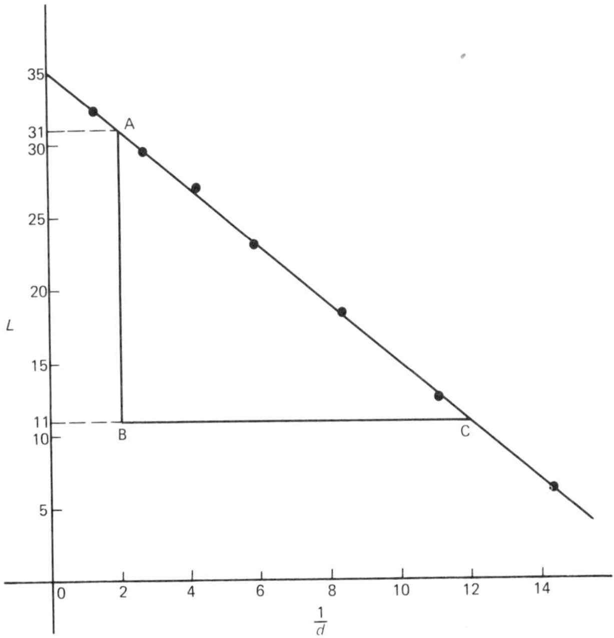

Fig 2

Gradient of straight line, $a = \frac{AB}{BC} = \frac{31 - 11}{2 - 12} = \frac{20}{-10} = -2$

L-axis intercept, b = **35**

Hence the law of the graph is $L = -\frac{2}{d} + 35$

When the distance d = 0.20 m, load $L = \frac{-2}{0.20} + 35$ = **25.0 N**

Rearranging $L = \frac{-2}{d} + 35$ gives $\frac{2}{d} = 35 - L$ and $d = \frac{2}{35 - L}$

Hence when load $L = 20$ N, distance, $d = \frac{2}{35 - 20} = \frac{2}{15} = \mathbf{0.133\ m}$

Problem 3 The following table gives corresponding values of two quantities x and y which are believed to be related by a law of the form $y = ax^2 + bx$, where a and b are constants.

x	5.0	9.4	15.6	21.0	29.5
y	57.5	141	312	510	917

By plotting a suitable graph, verify the law, and determine the approximate values of a and b. Hence determine (i) the value of y when x is 11.5, and (ii) the positive value of x when y is 292.5

Dividing both sides of $y = ax^2 + bx$ by x, gives

$$\frac{y}{x} = ax + b,$$

which is of the form $Y = m X + c$.

Thus $\frac{y}{x}$ is plotted vertically against x horizontally, and if a straight line results the law is verified. The gradient gives the constant 'a' and the vertical axis intercept gives the constant 'b'.
Another table of values is drawn up as shown below.

x	5.0	9.4	15.6	21.0	29.5
y	57.5	141	312	510	917
$\frac{y}{x}$	11.5	15.0	20.0	24.3	31.1

A graph of $\frac{y}{x}$ against x is shown in *Fig 3*. A straight line can be drawn through the points thus the law is verified.

Gradient of straight line, $a = \frac{AB}{BC} = \frac{23.6 - 15.6}{20 - 10} = \frac{8}{10} = \mathbf{0.8}$

Vertical axis intercept, $b = \mathbf{7.5}$

Hence the law of the graph is $y = \mathbf{0.8x^2 + 7.5x}$

(i) When $x = 11.5$, $y = 0.8(11.5)^2 + 7.5(11.5) = \mathbf{192.1}$

(ii) When $y = 292.5$, then $292.5 = 0.8x^2 + 7.5x$

i.e. $0.8x^2 + 7.5x - 292.5 = 0$

and $8x^2 + 75x - 2925 = 0$

Using the quadratic formula: $x = \frac{-75 \pm \sqrt{[(75)^2 - 4(8)(-2925)]}}{2(8)}$

$$= \frac{-75 \pm 315}{16} = \frac{240}{16} \text{ or } \frac{-390}{16}$$

$$= 15 \text{ or } -24.375$$

Hence the positive value of x **is 15** when $y = 292.5$

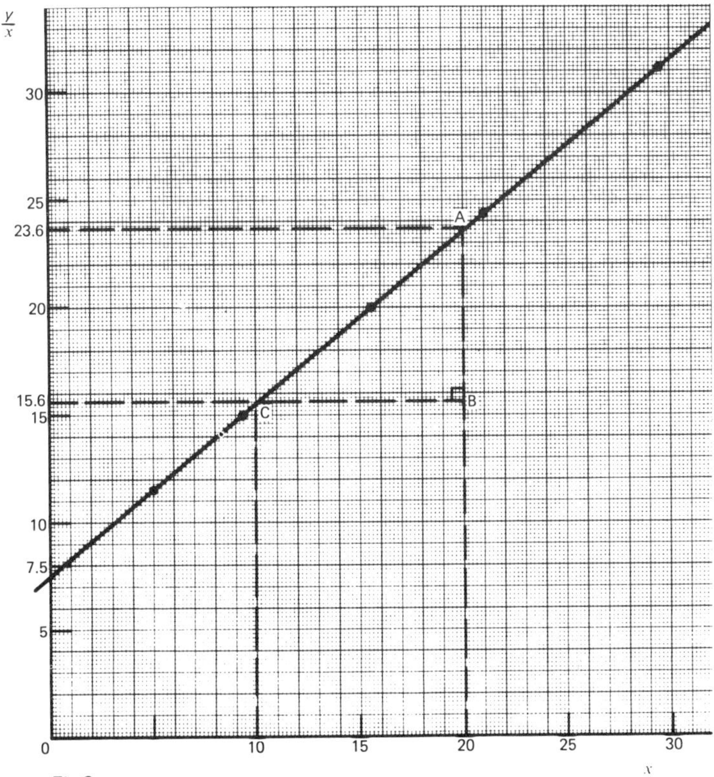

Fig 3

Problem 4 The solubility s of potassium chlorate is shown by the following table:

t°C	10	20	30	40	50	60	80	100
s	4.9	7.6	11.1	15.4	20.4	26.4	40.6	58.0

The relationship between s and t is thought to be of the form $s = 3 + at + bt^2$. Plot a graph to test the supposition and use the graph to find approximate values of a and b. Hence calculate the solubility of potassium chlorate at 70°C.

Rearranging $s = 3 + at + bt^2$ gives

$$s - 3 = at + bt^2$$

and

$$\frac{s-3}{t} = a + bt$$

or $\quad \frac{s-3}{t} = bt + a$, which is the

form $Y = mX + c$, showing that $\frac{s-3}{t}$ is to be plotted vertically and t horizontally. Another table of values is drawn up as shown below.

t	10	20	30	40	50	60	80	100
s	4.9	7.6	11.1	15.4	20.4	26.4	40.6	58.0
$\frac{s-3}{t}$	0.19	0.23	0.27	0.31	0.35	0.39	0.47	0.55

A graph of $\frac{s-3}{t}$ against t is shown plotted in *Fig 4*. A straight line fits the points which shows that s and t are related by $s = 3 + at + bt^2$.

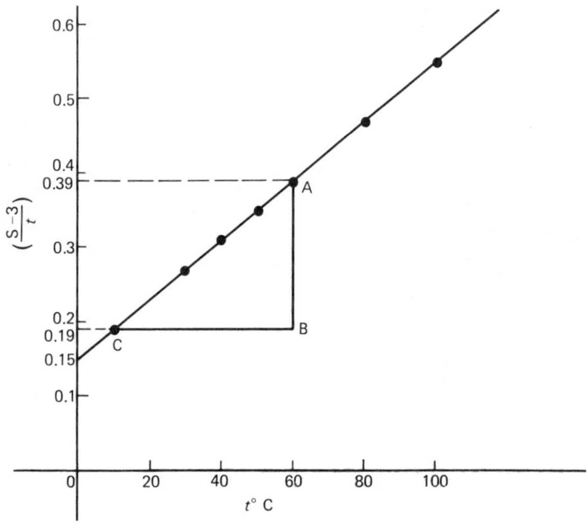

Fig 4

Gradient of straight line, $b = \frac{AB}{BC} = \frac{0.39 - 0.19}{60 - 10} = \frac{0.20}{50} = \mathbf{0.004}$

Vertical axis intercept, $a = \mathbf{0.15}$

Hence the law of the graph is $\mathbf{s = 3 + 0.15t + 0.004\, t^2}$

The solubility of potassium chlorate at 70°C is given by:

$s = 3 + 0.15(70) + 0.04(70)^2 = 3 + 10.5 + 19.6 = \mathbf{33.1}$

C. FURTHER PROBLEMS ON REDUCING NON-LINEAR LAWS TO LINEAR FORM

In *Problems 1 to 5*, x and y are two related variables and all other letters denote constants. For the stated laws to be verified it is necessary to plot graphs of the variables in a modified form. State for each (a) what should be plotted on the vertical axis, (b) what should be plotted on the horizontal axis, (c) the gradient and (d) the vertical axis intercept.

1 $y = d + cx^2$ [(a) y; (b) x^2; (c) c; (d) d]

2 $y - a = b\sqrt{x}$ [(a) y; (b) $\sqrt{x}$; (c) b; (d) a]

3 $y - e = \dfrac{f}{x}$ [(a) y; (b) $\dfrac{1}{x}$; (c) f; (d) e]

4 $y - cx = bx^2$ [(a) $\dfrac{y}{x}$; (b) x; (c) b; (d) c]

5 $y = \dfrac{a}{x} + bx$ [(a) $\dfrac{y}{x}$; (b) $\dfrac{1}{x^2}$; (c) a; (d) b]

6 In an experiment the resistance of wire is measured for wires of different diameters with the following results.

R ohms	1.64	1.14	0.89	0.76	0.63
d mm	1.10	1.42	1.75	2.04	2.56

It is thought that R is related to d by the law $R = \dfrac{a}{d^2} + b$, where a and b are constants. Verify this and find the approximate values for a and b. Determine the cross-sectional area needed for a resistance reading of 0.50 ohms.

[$a = 1.5$; $b = 0.4$; 11.78 mm^2]

7 Corresponding experimental values of two quantities x and y are given below.

x	1.5	3.0	4.5	6.0	7.5	9.0
y	11.5	25.0	47.5	79.0	119.5	169.0

By plotting a suitable graph verify that y and x are connected by a law of the form $y = kx^2 + c$, where k and c are constants. Determine the law of the graph and hence find the value of x when y is 60.0.

[$y = 2x^2 + 7$; 5.15]

8 Experimental results of the safe load, L kN, applied to girders of varying spans, d m, are shown below.

Span, d m	2.0	2.8	3.6	4.2	4.8
Load, L kN	475	339	264	226	198

It is believed that the relationship between load and span is $L = \dfrac{c}{d}$, where c is a constant. Determine (a) the value of constant c, and (b) the safe load for a span of 3.0 m.

[(a) 950; (b) 317 kN]

9 The following results give corresponding values of two quantities x and y which are believed to be related by a law of the form $y = ax^2 + bx$ where a and b are constants.

y	33.86	55.54	72.80	84.10	111.4	168.1
x	3.4	5.2	6.5	7.3	9.1	12.4

Verify the law and determine approximate values of a and b. Hence determine (i) the value of y when x is 8.0, and (ii) the value of x when y is 146.5.

[$a = 0.4$, $b = 8.6$; (i) 94.4; (ii) 11.2]

14 Simpson's rule and the theorems of Pappus

A. MAIN POINTS CONCERNED WITH SIMPSON'S RULE AND THEOREMS OF PAPPUS.

1 **Simpson's rule** provides a method whereby approximate areas of irregular plane surfaces may be determined. The method may be used, for example, by engineers estimating areas of indicator diagrams of steam engines, surveyors estimating areas of plots of land or naval architects estimating areas of water planes or transverse sections of ships.

2 To determine the area PQRS of *Fig. 1* using Simpson's rule:

(i) Divide base PS into an even number of intervals, each of width d, (the greater the number of intervals, the greater the accuracy).

(ii) Accurately measure ordinates y_1, y_2, y_3, etc.

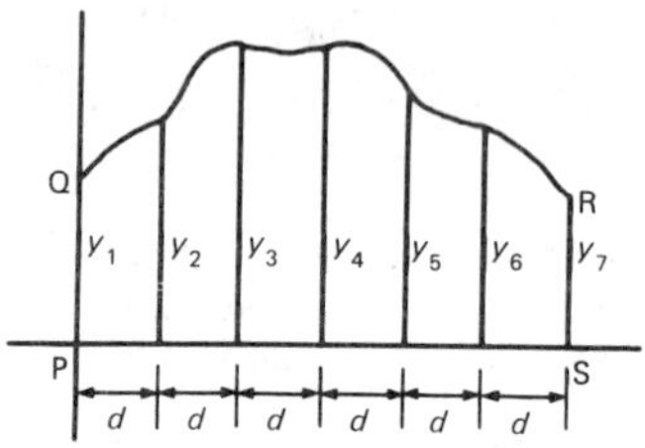

Fig 1

(iii) Area PQRS $= \frac{d}{3}[(y_1 + y_7) + 4(y_2 + y_4 + y_6) + 2(y_3 + y_5)]$

In general, Simpson's rule states:

$$\text{Area} = \frac{1}{3}(\text{width of interval})\left[\begin{matrix}\text{first + last}\\ \text{ordinate}\end{matrix} + 4\begin{pmatrix}\text{sum or even}\\ \text{ordinates}\end{pmatrix} + 2\begin{pmatrix}\text{sum of remaining}\\ \text{odd ordinates}\end{pmatrix}\right]$$

(See *Problems 1 to 3*)

3 Two theorems of Pappus enable surface areas and volumes to be calculated.

(a) The **first theorem of Pappus** states:

'If a plane curve is rotated about an axis in its own plane but not intersecting it, the surface area generated is given by the product of the arc length and the distance moved by the centroid of the curve.'

i.e. Surface area generated = arc length × distance moved through by the centroid.

In *Fig 2*, if the curve of length s is rotated one revolution about the x-axis and C is the position of the centroid of the curve then the distance moved by the centroid when the curve revolves one revolution about OX is $2\pi\bar{y}$ (i.e., the circumference of a circle). Hence by Pappus' theorem, **surface area** = arc length × $2\pi\bar{y}$

$= 2\pi s\,\bar{y}$ **square units.**

(b) The **second theorem of Pappus** states:
'If a plane area is rotated about an axis in its own plane but not intersecting it, the volume of the solid formed is given by the product of the area and the distance moved by the centroid of the area.'
i.e. Volume generated = area × distance moved through by the centroid. In *Fig 3*, let C be the centroid of area A, and let $\bar{y}$ be the perpendicular distance of C from axis OX, then the distance moved by the centroid when area A makes one revolution about OX is $2\pi \bar{y}$ (i.e. the circumference of a circle). Hence by Pappus' theorem, Volume generated = area × $2\pi\bar{y}$
i.e. $V = 2\pi A \bar{y}$ **cubic units.**
(See *Problems 4 to 8*)

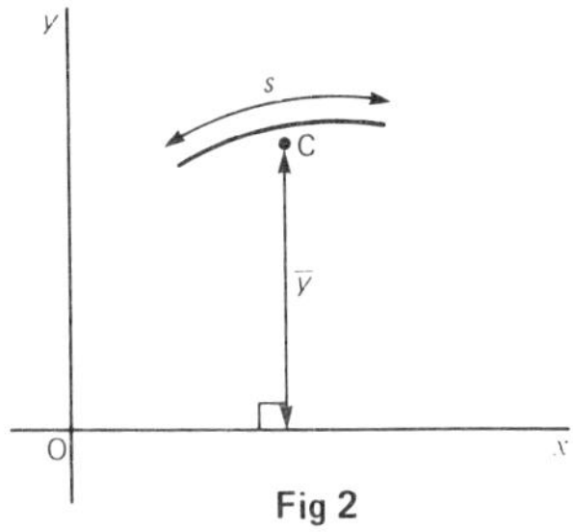

Fig 2

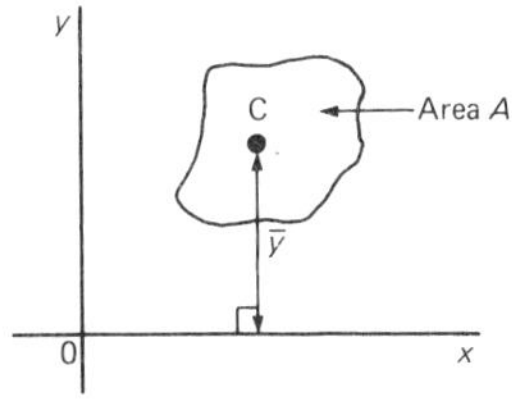

Fig 3

B. WORKED PROBLEMS ON SIMPSON'S RULE AND THE THEOREMS OF PAPPUS

Problem 1 A car starts from rest and its speed is measured every second for 6 seconds.

Time t (s)	0	1	2	3	4	5	6
Speed v (m/s)	0	2.5	5.5	8.75	12.5	17.5	24.0

Determine the distance travelled in 6 s (i.e., the area under the v/t graph) by using Simpson's rule.

A graph of speed/time is shown in *Fig 4*.

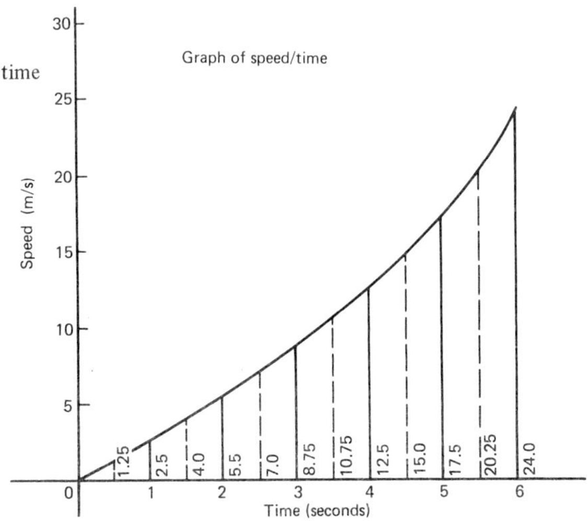

Fig 4

Let the time base be divided into 6 strips each of width 1 second, and the length of the ordinates measured. Then from para. 2,

$$\text{area} = \frac{1}{3}(1)\,[(0 + 24.0) + 4(2.5 + 8.75 + 17.5) + 2(5.5 + 12.5)] = \mathbf{58.33\ m}$$

With Simpson's rule the greater the number of intervals chosen, the greater is the accuracy of the result obtained. In this case, if the time base is divided into 12 strips each of width 0.5 s, and the length of the ordinates measured then

$$\text{area} = \frac{1}{3}(0.5)\left[\begin{matrix}(0 + 24.0) + 4(1.25 + 4.0 + 7.0 + 10.75 + 15.0 + 20.25)\\ + 2(2.5 + 5.5 + 8.75 + 12.5 + 17.5)\end{matrix}\right]$$

$$= \frac{1}{3}(0.5)\,[24.0 + 233 + 93.5]$$

$$= 58.42\ \text{m}$$

Hence the distance travelled in 6 seconds is 58.42 m.

Problem 2 A river is 15 m wide. Soundings of the depth are made at equal intervals of 3 m across the river and are as shown below.

Depth (m)	0	2.2	3.3	4.5	4.2	2.4	0

Calculate the cross-sectional area of the flow of water at this point using Simpson's rule.

$$\text{From para. 2, Area} = \frac{1}{3}(3)\,[(0 + 0) + 4(2.2 + 4.5 + 2.4) + 2(3.3 + 4.2)]$$

$$= (1)\,[0 + 36.4 + 15] = \mathbf{51.4\ m^2}$$

Problem 3 An indicator diagram for a steam engine is shown in *Fig 5*. The base line has been divided into 6 equally spaced intervals and the lengths of the 7 ordinates measured with the results shown in centimetres. Determine (a) the area of the indicator diagram using Simpson's rule, and (b) the mean pressure in the cylinder given that 1 cm represents 100 kPa and mean pressure $= \dfrac{\text{area of diagram}}{\text{length of base.}}$

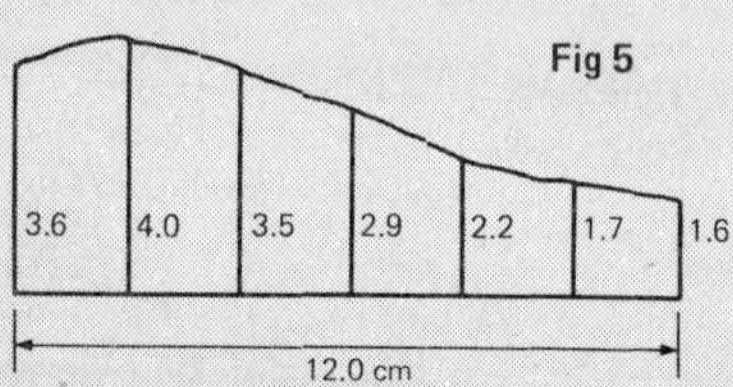

Fig 5

(a) The width of each interval is $\dfrac{12.0}{6} = 2.0$ cm

$$\text{Using Simpson's rule, area} = \frac{1}{3}(2.0)\left[\begin{matrix}(3.6 + 1.6) + 4(4.0 + 2.9 + 1.7\\ + 2(3.5 + 2.2)\end{matrix}\right]$$

$$= \frac{2}{3}\,[5.2 + 34.4 + 11.4] \quad = \mathbf{34\ cm^2}$$

(b) Mean height of ordinates $= \dfrac{\text{area of diagram}}{\text{length of base}} = \dfrac{34}{12} = \mathbf{2.83\ cm.}$

Since 1 cm represents 100 kPa, the mean pressure in the cylinder

$$= 2.83\ \text{cm} \times 100\ \frac{\text{kPa}}{\text{cm}}$$

$$= \mathbf{283\ kPa.}$$

Problem 4 Determine (a) the surface area and (b) the volume of an anchor ring formed by rotating a circle of radius 6.0 cm about an axis at a distance of 25 cm from its centre. Give answers correct to 3 significant figures.

The 6.0 cm radius circle is rotated about axis XX as shown by the side elevation in *Fig 6*.

(a) By the first theorem of Pappus, (see para. 3(a)), surface area = arc length × distance moved by centroid.

Arc length = circumference of 6.0 cm radius circle = $2\pi(6.0) = 12\pi$ cm.

Distance moved by centroid C in one complete revolution = $2\pi(25) = 50\pi$ cm, (i.e., circumference of a circle of radius 25 cm).

Hence **surface area of anchor ring** = $(12\pi)(50\pi)$ = **5920 cm²**, correct to 3 significant figures.

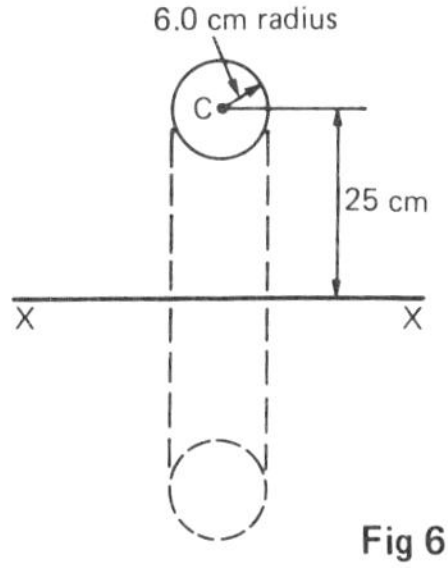

Fig 6

(b) By the second theorem of Pappus, (see para. 3(b)), volume generated = area × distance moved by centroid

Area of circle = $\pi(6.0)^2 = 36.0\pi$ cm^2

Hence **volume of anchor ring** = $(36.0\pi)(50\pi) = 17\,770$ cm^3 = $\dfrac{17\,770}{10^6}$ m^3

= **0.0178 m³**, correct to 3 significant figures

Problem 5 Use the theorem of Pappus to calculate the position of the centroid of a semicircular area of radius *r*.

If the semicircular area shown in *Fig 7* is rotated about axis XX, then the volume of the solid generated is that of a sphere of volume $\frac{4}{3}\pi r^3$. The area of the semicircle is $\frac{\pi r^2}{2}$. Centroid C is shown in *Fig 7* on the axis of symmetry OY. By the theorem of Pappus, volume generated = area × distance moved through by centroid

i.e. $\frac{4}{3}\pi r^3 = \frac{\pi r^2}{2}(2\pi\bar{y})$

Hence $\bar{y} = \dfrac{\frac{4}{3}\pi r^3}{(\frac{\pi r^2}{2})(2\pi)} = \dfrac{4r}{3\pi}$ units

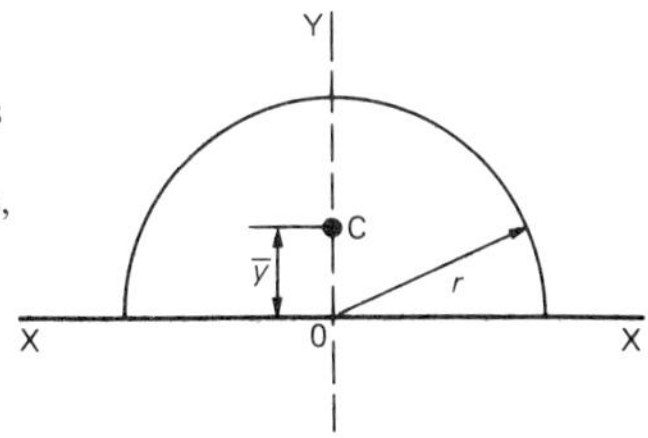

Fig 7

i.e. the centroid of a semicircular area lies on the axis of symmetry at a distance of $\frac{4r}{3\pi}$ from the diameter.

Problem 6 Determine the position of the centroid of a semicircular arc of radius r using the theorem of Pappus

If the curve shown in *Fig 8* is rotated one revolution about axis XX the surface area generated is the surface area of a sphere, i.e., $4\pi r^2$. C is the position of the centroid, perpendicular distance $\bar{y}$ from axis XX. By the first theorem of Pappus,
surface area = arc length $\times\ 2\pi\bar{y}$.

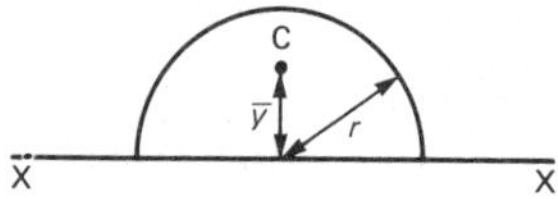

Fig 8

The arc length = half the circumference of the circle of radius r, i.e. πr.

Hence $\quad 4\pi r^2 = \pi r \times 2\pi\bar{y}$

from which, $\quad \bar{y} = \dfrac{4\pi r^2}{2\pi^2 r} = \dfrac{2r}{\pi}$

Hence the centroid of a semicircular arc is $\dfrac{2r}{\pi}$ from the axis of rotation on the axis of symmetry.

Problem 7 A cylindrical pillar of diameter 400 mm has a groove cut round its circumference. The section of the groove is a semicircle of diameter 50 mm. Determine the volume of material removed, in cubic centimetres, correct to 4 significant figures.

A part of the pillar showing the groove is shown in *Fig 9*. The distance of the centroid of the semicircle from its base is $\dfrac{4r}{3\pi}$ (see *Problem 5*) = $\dfrac{4(25)}{3\pi}$

$= \dfrac{100}{3\pi}$ mm.

The distance of the centroid from the centre of the pillar

$= (200 - \dfrac{100}{3\pi})$ mm.

The distance moved through by the centroid in one revolution

$= 2\pi(200 - \dfrac{100}{3\pi})$

$= (400\pi - \dfrac{200}{3})$ mm.

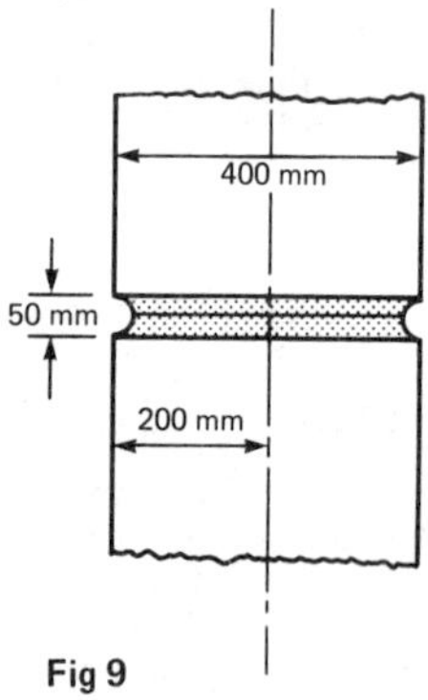

Fig 9

From the second theorem of Pappus, volume = area $\times$ distance moved by centroid

$$= (\frac{1}{2}\pi 25^2)(400\pi - \frac{200}{3}) = 1\ 168\ 250\ \text{mm}^3$$

Hence the volume of material removed = 1168 cm³, correct to 4 significant figures.

Problem 8 A metal disc has a radius of 5.0 cm and is of thickness 2.0 cm. A semi-circular groove of diameter 2.0 cm is machined centrally around the rim to form a pulley. Determine, using Pappus' theorem, the volume and mass of metal removed and the volume and mass of the pulley if the density of the metal is 8000 kg m^{-3}.

A side view of the rim of the disc is shown in *Fig 10*. When area PQRS is rotated about axis XX the volume generated is that of the pulley. The centroid of the semicircular area removed is at a distance $4r/3\pi$ from its diameter (see *Problem 5*), i.e. $4(1.0)/3\pi$, i.e. 0.424 cm from PQ. Thus the distance of the centroid from XX is 5.0 – 0.424, i.e., 4.576 cm. The distance moved through in one revolution by the centroid is $2\pi(4.576)$ cm.

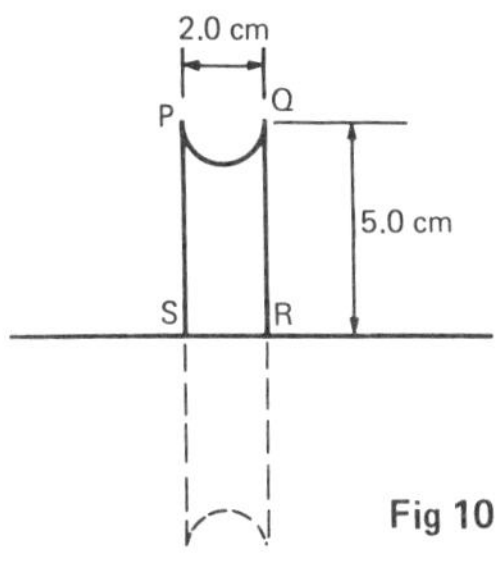

Fig 10

Area of semicircle $= \dfrac{\pi r^2}{2} = \dfrac{\pi(1.0)^2}{2} = \dfrac{\pi}{2}$ cm^2.

By the second theorem of Pappus, volume generated = area × distance moved by centroid $= (\dfrac{\pi}{2})\,(2\pi\ 4.576)$

i.e. **volume of metal removed = 45.16 cm^3**.

Mass of metal removed = density × volume

$= 8000 \text{ kg m}^{-3} \times \dfrac{45.16}{10^6} \text{ m}^3 =$ **0.3613 kg or 361.3 g.**

Volume of pulley = volume of cylindrical disc – volume of metal removed
$= \pi\,(5.0)^2\ (2.0) - 45.16 =$ **111.9 cm^3**

Mass of pulley = density × volume

$= 8000 \text{ kg m}^{-3} \times \dfrac{111.9}{10^6} \text{ m}^3$

= **0.8952 kg or 895.2 g.**

C. FURTHER PROBLEMS ON SIMPSON'S RULE AND THE THEOREMS OF PAPPUS

SIMPSONS' RULE

1 Plot a graph of $y = 3x - x^2$ by compiling a table of values of y from $x = 0$ to $x = 3$. Determine the area enclosed by the curve, the x-axis and ordinate $x = 0$ and $x = 3$ by using Simpson's rule. [4½ square units]

2 Plot the graph of $y = 2x^2 + 3$ between $x = 0$ and $x = 4$. Estimate the area enclosed by the curve, the ordinates $x = 0$ and $x = 4$, and the x-axis by using Simpson's rule. [54.7 square units]

3 The velocities of a car at one second intervals are given in the following table:

time t(s)	0	1	2	3	4	5	6
velocity v (m/s)	0	2.0	4.5	8.0	14.0	21.0	29.0

Determine the distance travelled in 6 seconds (i.e. the area under the v/t graph) using an approximate method. [63 m]

4 The shape of a piece of land is shown in *Fig 11*. To estimate the area of the land, a surveyor takes measurements at intervals of 50 m, perpendicular to the straight portion with the results shown, (the dimensions being in metres). Estimate the area of the land in hectares (1 ha = 10^4 m^2), using Simpson's rule. [4.70 ha]

Fig 11

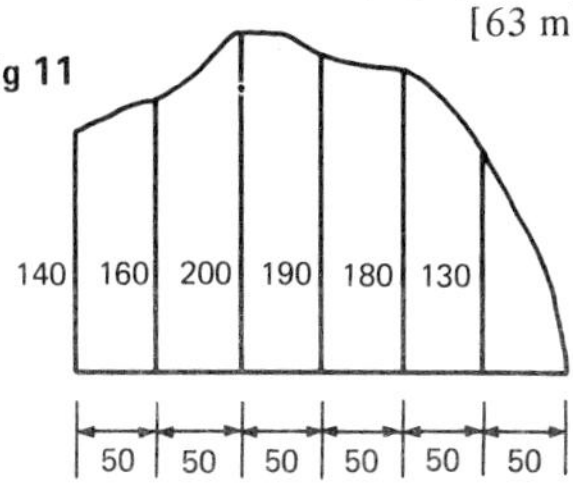

5 The deck of a ship is 35 m long. At equal intervals of 5 m the width is given by the following table:

Width (m)	0	2.8	5.2	6.5	5.8	4.1	3.0	2.3

Estimate the area of the deck.

[143 m^2]

6 An indicator diagram of a steam engine is 12 cm long. Seven evenly spaced ordinates, including the end ordinates, are measured as follows: 5.90, 5.52, 4.22, 3.63, 3.32, 3.24, 3.16 cm.
Determine the area of the diagram using Simpson's rule and the mean pressure in the cylinder if 1 cm represents 90 kPa, given that mean pressure $= \dfrac{\text{area of diagram}}{\text{length of base}}$.

[49.13 cm^2 : 368.5 kPa]

THEOREMS OF PAPPUS

7 A right angled isosceles triangle having a hypotenuse of 8 cm is revolved one revolution about one of its equal sides as axis. Determine the volume of the solid generated using Pappus' theorem.

[189.6 cm^3]

8 A rectangle measuring 10.0 cm by 6.0 cm rotates one revolution about one of its longest sides as axis. Determine the volume of the resulting cylinder by using the theorem of Pappus.

[1131 cm^2]

9 Determine the surface area and the volume of a metal ring formed by rotating a circle of radius 9.0 cm about an axis at a perpendicular distance of 20.0 cm from its centre.

[7106 cm^2; 31 980 cm^3]

10 Show by using the theorem of Pappus that the centroid of a semicircular arc of radius 50 mm lies on the axis of symmetry 18.17 mm from the arc.

11 A metal disc has a radius of 7.0 cm and is of thickness 2.5 cm. A semicircular groove of diameter 2.0 cm is machined centrally around the rim to form a pulley. Determine the volume of metal removed using Pappus' theorem and express this as a percentage of the original volume of the disc. Find also the mass of metal removed if the density of the metal is 7800 kg m^{-3}.

[64.90 cm^3; 16.86%; 506.2 g]

12 One edge of an equilateral triangle of side 10.0 cm is parallel to an axis XX lying at a distance of 12.0 cm from the edge and the apex is on the opposite side to the axis. If the triangle is rotated 360° about axis XX, determine the surface area and the volume of the solid generated.

[2806 cm^2; 4050 cm^3]

Answers to multi-choice problems

Chapter 4 (page 64)

1 (b); 2 (e); 3 (e); 4 (c); 5 (c); 6 (a); 7 (d); 8 (f); 9 (b); 10 (c); 11 (b); 12 (a); 13 (b); 14 (b).

Chapter 5 (page 77)

1 (c); 2 (d); 3 (d); 4 (a); 5 (d); 6 (c); 7 (b); 8 (c); 9 (b); 10 (c).

Chapter 6 (page 87)

1 (c); 2 (a); 3 (b); 4 (b); 5 (a); 6 (d); 7 (d); 8 (b); 9 (c); 10 (b).

Index